LIFE IN THE BARN

KYLE BOOKS

An Hachette UK Company
www.hachette.co.uk

First published in Great Britain in 2022 by
Kyle Books, an imprint of Octopus Publishing Group Limited
Carmelite House
50 Victoria Embankment
London EC4Y 0DZ
www.kylebooks.co.uk

This edition published in 2022

ISBN: 978 1 91423 964 9

Distributed in the US by Hachette Book Group, 1290 Avenue of the Americas,
4th and 5th Floors, New York, NY 10104

Distributed in Canada by Canadian Manda Group, 664 Annette St., Toronto,
Ontario, Canada M6S 2C8

Publisher: Joanna Copestick
Editorial Assistant: Emma Hanson
Design: Rachel Cross
Photography: Sarah Weal
Production: Peter Hunt

Text by: Jayne Dowle and Elizabeth Wilhide

Printed in Italy

10 9 8 7 6 5 4 3

Contents

Foreword

Jay Blades

AT THE REPAIR SHOP WE'RE LIKE A FAMILY. And I'm certain that this closeness has come about because we were allowed to grow as a team almost organically: we aren't a group of people who have been forced together. Instead, what we all have in common are our skills and love of craft.

When you work with someone six days a week for five years, you really get to know one another. You learn that person's concerns, their family woes. You get to know each other's strengths and weaknesses. We like taking the mickey out of each other, like normal families do. We have an unbelievable laugh and lots of banter. The barn is a very cosy place to be. It's almost as if you go to work, leaving one family behind, and then when you step inside the barn, you're welcomed into another one.

It's not just the team in front of the camera, but also the people behind it. I'm in awe of them. They do the scheduling, they do the filming and editing, and they do so much to make sure that what viewers see on TV is seamless, and that takes great work.

Some of the objects that come into the barn don't look very prepossessing when you first set eyes on them. Some are so battered and worn that in other circumstances you and I would probably throw them away. But you have to hear the story to make sense of the item. It might be a domino set made by a grandfather during World War I and played in the famous 1914 Christmas Truce, or a handwritten recipe book handed down the generations and literally falling apart at the seams. Very few of the items are important enough to go into a museum, but they're strongly linked to that particular family. They're social history.

The true value of these objects is not monetary at all, it's emotional. You would never say to someone how much would you give me for my memories? They're yours, they're personal to you. It's a ridiculous question to ask. For a family to entrust us with an item that has so much meaning for them is very moving. To reunite someone with their past, to give a beloved item a future so that it can be handed down the succeeding generations, is incredibly humbling.

When people ask me if I have a favourite item, I usually say it's like picking your favourite child. You can't do it, they're all special. But I must admit that the stories that move me a lot are the ones when men get emotional. Two examples spring to mind. There was a guy called Albert, who brought in a radio that had stopped working when his wife passed away, and another guy called

Geoff, who had been unable to listen to 'Moonlight Serenade' on his jukebox because it was the first dance he and his wife had had at their wedding reception.

Where I was brought up, men were supposed to be tough. They weren't supposed to show their vulnerability. So when I see men come into the barn, openly expressing their love for their late partners, it makes me so proud that they are prepared to show their emotions on national TV. As a society, that's something we need to address.

When you break it down, I think the whole show is about sustainability. In an increasingly disposable age, we're conveying an important message about how important it is to repair something, not simply throw it away. But sustainability isn't just about repair and restoration, it isn't just about separating your plastics from your foil and your paper and stuff like that, it's about people. If you're able to repair an item that brings back memories for someone, you're making sure they can keep that memory alive. In the barn, we not only repair items, we repair people and their family histories.

As many people know, I have dyslexia and never learned to read properly – I recently made a documentary about what it's like to learn to read at the age of 50. But my dyslexia doesn't really affect my work at The Repair Shop – in fact, it's one of my biggest superpowers. Because I don't read that well, I tend not to like to have the information about an item or a family beforehand.

About five minutes before someone arrives, I'll find out their name, then I'm free to have a proper conversation with them on the spot. Why is the item important to them? How did it get broken? What are their hopes for its future? You show a true reaction when you learn about someone's history for the first time. If you've read all about it before, your reaction might be a bit tame. I always prefer the human way, just talking to people.

It's the same with craft. These days, with everything going digital and most people spending their days in front of a screen, it's easy to forget what computers can't do. Making something or fixing something by hand is a deeply satisfying way of working. Nothing can beat the human touch. People really get into the zone – every craftsperson has one of those. For me, it's sanding. I love repairing a piece of furniture, it's almost like it's talking to you. I also like ironing, but I'm quite weird like that!

When I first ran a charity years ago, teaching young offenders how to restore and repair old furniture, I knew next to nothing about craft. Luckily, the charity was based in High Wycombe, which used to be the furniture capital of Britain. So I went to Age Concern, the Women's Institute and Neighbourhood Watch, and asked their members if they would come and teach us their skills. We were inundated with offers. Our oldest teacher was a 92-year-old who taught us how to cane and rush a chair.

My present furniture restoration business is based in Ironbridge, where we repair items, carry out commissions and redesign second-hand furniture to give it a new lease of life and stop it going to landfill. Nowadays, filming means I don't have much time to get my hands dirty, I've become more of a designer and less a designer-maker – someone else does the hands-on work for me. It's the same at The Repair Shop. When I joined the show, I was doing the upholstery for the first two series until it became clear that I wasn't going to get the work done if I was hosting as well. But I get a great deal of pleasure and satisfaction watching the others do their repairs. There's always something new to learn.

'Collaboration is about teamwork, which is the best way to move forward in the world of craft. It's also about watching and learning at the same time. You often pick up a technique from another expert that lends itself to your own discipline quite easily.'

THE MAGIC OF THE BARN

A WARM WELCOME AND EXPERT
CRAFTSPEOPLE AT WORK

Making the magic

Ever since the first series of *The Repair Shop* aired in 2017, it has been filmed at the Weald & Downland Living Museum in West Sussex. Set in the rolling countryside of the South Downs, this open-air museum is home to a collection of historical farm buildings, relocated from all over the country. Court Barn, the programme's principal workshop, is a thatched, timber-framed structure dating from the late 17th to early 18th century; it was originally sited on a Hampshire farm. For a show devoted to repairing and restoring precious family treasures, it seems especially fitting that this ancient building has found a whole new lease of life as the home of *The Repair Shop*'s team of dedicated experts.

If there's a make-do-and-mend quality to the barn interior, it's entirely appropriate. A number of the workbenches and lights were made by metalworker Dom Chinea, who was also responsible for the show's iconic sign that greets visitors as they approach the door. The experts' tools of trade are everywhere, neatly arranged and readily to hand – from Steve Fletcher's impressive collection of pliers to Will Kirk's cherished mallet made of olive wood. If the occasional bird flutters in from the surrounding fields to roost under the thatch, it's all part of the charm.

But the true magic of the barn arises from something much less tangible. Just as you could never put a price tag on a beloved old doll that has seen better days, or on a seized-up go-kart someone's grandad cobbled together from whatever was lying around in his shed, the magic of the barn lies in the power of memories. A side-saddle that takes a woman back to her younger days and her Wild West adventures, an Omani chest studded with brass nails that stands for a vanished life in a different country, a 1950s radiogram that once provided the soundtrack to family Sundays – such objects not only bring the past into sharp focus, they are held in the deepest affection, which has nothing to do with their financial value.

If their owners understand this instinctively, so do the experts. They know that when they repair one of these humble treasures, it goes far beyond the cosmetic or the merely functional. Each restoration is like putting a missing piece of a family jigsaw back into place and, when it's time for handover, they are often as moved as the delighted recipients.

Over the years a real camaraderie has grown up in the barn and this, too, is part of the magic. It's obvious how willing the experts are to pick up new techniques from each other, to pool their skills and collaborate, and to puzzle out just how to fix an object that they might never have come across in their working lives before. They all learn from each other. As Jay says, there must be 600 years of combined expertise in the barn. It's fair to say that there is nowhere else on earth like this very special place.

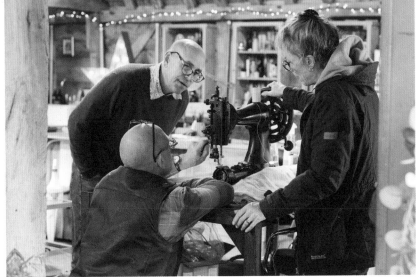

What makes a great repair?

A great repair starts with listening. Every object that comes into the barn has its own history. Finding out what makes it special to its owner and the hopes they have for its repair is always the first step. There may be a name written or inscribed on it somewhere, for example, that they want to preserve, or some other marker of its previous use and association.

In the case of mechanical items that no longer function, returning them in working order is the bottom line. Sometimes this demands plenty of elbow grease to disassemble rusted parts and scrape off the evidence of years of neglect. In the case of electronic devices, which might have gone well beyond their expected lifetime of service, it may be necessary not only to fix what's broken, but also to substitute key elements that may fail sooner rather than later, so that the items carry on functioning long after they have been handed back.

There's a fine line between repair and conservation. Judging how far to go is an important part of the process, which means being sensitive to the object's past and the signs of wear that are part of its story. While many of the transformations are stunning and dramatic, obliterating all the nicks and scratches that convey a history of use is not the point. The Teddy Bear Ladies, for example, are happy to restore battered soft toys to their former plump selves with new stuffing, and put smiles back on their faces, but the bald patches where fur has been worn away are always left as testimony of years of being hugged and loved.

The experts at The Repair Shop like a challenge. Handmade or one-off items demand an element of reverse engineering or detective work, to get into the mindset of the original maker or puzzle out exactly how everything fits together. Occasionally, it's a steep learning curve, which makes the results all the more satisfying.

Finally, great repairs take time. What viewers see is often the tip of the iceberg. Handmaking 80 brass nails doesn't happen in a morning. But even routine steps such as cleaning, sanding and polishing demand patience and meticulous attention to detail, which may take many hours, if not days.

Why craft matters

Craft is a living heritage that connects the present with years gone by. But there's no getting away from the fact that over the past few decades our throwaway culture and addiction to cheap mass-produced goods have seen such time-honoured practices wane. The make-do-and-mend ethos of the war years was successful because most people had at least a few skills to call upon, whether it was carpentry, darning or basic metalwork. More often than not, such expertise was handed down the generations, picked up from parents and grandparents. Today, this is often no longer the case.

Advances in technology have played an obvious role. When you no longer rely solely on horsepower for farming, transport or to get from place to place, it's obvious that you aren't going to need so many saddlers, harness-makers or blacksmiths. Then there's the fact that many modern products, especially appliances and cars, have digital components that are not designed to be repaired by the average person, which inevitably gives them a reduced lifespan.

But there are encouraging signs that craft is making a comeback. When so much work and leisure happens virtually, at one remove on a screen, there's been a renewed hunger for what is tangible and handmade. Knitting and pottery classes, for example, are heavily oversubscribed, and a new generation is rediscovering the warm and immediate aural quality of valve radios and vinyl records.

Working with your hands is centring and grounding: the hand-eye-brain connection is very meditative. For upholsterer Sonnaz Nooranvary, slip-stitching is her happy place. In a similar way, craft entails getting acquainted with tools and their working intelligence. It's no real surprise that for many of the experts in the barn, old tools are the best, not simply because they are often better made than modern versions, but also because they offer a satisfying sense of continuity. Unlike some of the experts in the barn, metalworker Dom Chinea did not inherit any of his tools. Instead, he prizes the ones he's picked up over the years at car boot sales because he knows they once belonged to somebody who used and cherished them.

Most importantly, at a time of climate change, craft is the route to repair, which means it's good for the planet. As mechanical whizz Mark Stuckey puts it, the greenest thing you can do is keep your old things in working order.

THE
EXPERTS

the
metalworker

Dominic Chinea

Regular viewers of *The Repair Shop* will be more than familiar with the iconic sign that welcomes visitors to the barn where all the magic happens. What few will realize, however, is that it's the handiwork of one of the programme's resident experts, metalworker Dom Chinea.

Dom's roundabout route to The Repair Shop began with a teenage fascination for skateboards, pushbikes, BMXs and old cars – anything with wheels. Always practical and hands-on, he taught himself to weld before he had even passed his driving test. From his first car, an old air-cooled VW Beetle, he graduated to camper vans, travelling to Cornwall on holiday with friends, which inevitably meant tinkering by the side of the road when an engine broke down.

Cars – especially VWs – remained a constant theme throughout Dom's time at college. While he was studying for a degree in graphic design at the University of Essex, he took a part-time job at Karmann Konnection in Southend, a well-known VW restorer, where he was exposed to a broad range of different repairing techniques. Then, in a slight departure, he went to work for a Southend commercial set-building firm called Rocket Art, where he added more skills to his repertoire.

Dom's roundabout route to The Repair
Shop began with a teenage fascination
for skateboards, pushbikes, BMXs and old
cars – anything with wheels.

Before long, Dom had set his sights on working with leading celebrity and fashion photographer Rankin, also renowned as the co-founder of the magazine *Dazed & Confused*. Rankin regularly offered two unpaid three-month internships, which were heavily subscribed. It took a lot of persistence from Dom, but after eight months he eventually secured one of these sought-after positions, sold his car and moved to London. At the end of the internship, he was offered a job.

Following an intense period of five or six years. Dom progressed from photographer's assistant to set designer and builder. During this time, he travelled the world on assignments with Rankin, and was asked to turn his hand to everything from lighting to spur-of-the-moment prop-finding. Although Rankin did not employ a set designer, Dom gradually assumed this role himself, building up his own workshop in a cupboard in the corner of a courtyard. Eventually, with the aid of an assistant, he was constructing full-scale sets.

Although the work was stimulating and enjoyable, the pace was relentless. One day Dom decided to take the plunge and set up on his own. It was an enormous step and a pivotal point in his career, but one that was to pay off handsomely. Over the following years, he assembled his own list of clients, and relished the multidisciplinary nature of the work, designing sets, constructing and painting them, and finding the props, often to a very tight deadline. By this time he had moved out of London and established a bigger workshop in Kent.

When Ricochet TV approached Dom to make the programme's sign, it initially seemed like just another job. Yet the producers had already spotted a short film he'd posted on YouTube and were keen to interview him for the show. After he'd designed and created the sign, he made some workbenches, and painted signs and other bits and pieces for the barn interior, too. Before long, he was being asked back more and more often to work on projects of his own.

I like doing practical things, putting things back into use, and getting them working again. I'm not so interested in what's purely ornamental or decorative. If it's an old butter churn and we can repair it so it makes butter again, or a hay press that we can get to function as it used to, I find that very satisfying. One of my favourite projects at the barn was the bright red Austin J40 pedal car – a beautiful, elaborate object. I'm very lucky to work on such a wide variety of things.

At The Repair Shop, I probably tend to work in collaboration more than the other experts. One of them will take something apart and realize that a bracket is broken and they will need me to weld or braze it back together. Or someone else will ask me to spray-paint something they're working on. Although I'm described as a metalworker, I've never liked being pigeonholed – it all goes back to my days at Karmann Konnection, when I was just as interested in the woodwork and upholstery side of car restoration as I was in the welding and panel-beating.

At the same time, I do have my own jobs and these tend to be among the bigger items and often take the greatest amount of time. Some people might think it's daft to put so much effort into what will end up as only eight minutes of TV, but I really enjoy the challenge and I know that many of these items wouldn't get repaired if it wasn't for this show. It just wouldn't be cost-effective. Those of us who work at the barn want to do what's best for the item or for the owners who brought it in. No one's looking for the cheapest or easiest way out.

Jobs can get bigger quickly, especially if you are repairing a previous repair, and what should be a simple case of disassembly, rebuilding and repainting turns out to be much more involved because parts are missing. If a nut is missing, for example, then you may have to measure the threads to work out how to make a replacement.

In the case of a homemade object, you have to get into the mind of the person who made it and follow their thought processes. I once had to repair a telescope that the owner's grandfather had made during the 1940s. At the time, the grandfather worked for the water board and all the nuts, bolts and levers were from water pipes. The tube of the telescope was fashioned out of a wooden shop roller and the end was a section of a big pipe that he'd cut off. The bearings, which probably came from a kid's toy, were missing. These projects can be difficult to do but they're very satisfying. An object that someone has handmade for their kids can be very charming and you can appreciate all the problem-solving that went into it.

Another handmade item that I worked on was a child's jeep, made during the war by a father who was building decoys for the military with a team of engineers. It was made of wood, had two speeds and a little gear stick. Once I'd disassembled it, what was tricky about the repair was working out the order in which it went back together. Commercially made machinery, such as farm equipment, is designed to be taken apart, in the middle of a field if necessary, and reassembled on the spot. With the jeep, the original builders had screwed bits together in the order they made them, so I had to work out how to rebuild it the same way.

Although I do have modern tools, such as some sockets and spanners, mostly I work with old ones. This goes back to the days when I was a kid scouring car boot sales for bargains, with only the money from my paper round to spend. None of my tools belonged to my grandfather or anything romantic like that, but they were someone's grandad's old tools. You appreciate who used them before you, and they are often made better and have worn well, which is why it's important to look after them. They're tactile things and you have an intimate relationship with them. Nowadays I still enjoy going to car boot sales – partly because it's the thrill of the chase.

My work often involves signwriting, lettering or painting. I tend to keep everything in a selection of engineer's boxes, filling up the drawers with pigments and brushes.

But I don't have a favourite tool: it changes every day. It might be a screwdriver that turns out to be just the right length to get into an awkward place. Or a vice if I'm trying to undo a really stubborn bolt. At those tense times, the tool becomes something you rely on and trust. It's your friend for the job.

The Repair Shop is such an inspiring place to be. We all learn from each other. I do get nervous when I stand at that reveals table – I've always been overcritical of my work. But when I can stand back and return an item to its owner, I feel proud knowing that I've done the best I could.

Having spent 10–15 years on the other side of the camera, holding up a set in the background or dangling over a ladder with a chandelier, I never for a second thought I would end up on the other side of it!

Pedal Train

Full steam ahead

Ask Kate Humphries about her favourite childhood memory and immediately she will say it was playing in the garden on the bright red and green pedal train her father, Barry, had bought in the early 1960s.

Although it was more than 30 years old by the time Kate got to play with it in the 1980s , the vintage train, constructed from plywood and metal, was a huge favourite. Barry, an avid train fan, loved it almost as much as all of his children did.

Sadly, now it has reached the latest generation – Kate's daughter, Agnes, is two and she has a baby brother, Euros – the train is in such a dilapidated state of repair it's hardly possible to play on it. It's covered in rust and looks very sorry for itself. And Barry, whose twin passions – trains and family – were epitomised in this ride-on toy, had a stroke in March 2021. His daughter, who lives in

Wales, thinks that seeing his grandchildren enjoying themselves with his beloved train will give her dad just the lift he needs.

Trains were always his passion, Kate says: "the sound, the smell, the history". He loved to explore the train graveyard in Barry Island, South Wales, and take his family to the Launceston Steam Railway in Cornwall, riding on the engine while Kate and her sisters stayed in the carriage.

Barry was always so keen to help other people and now Kate thinks that it's time that he – and the train he bought all those years ago – were looked after in return. She would love for the train to be safe once again for her own children and her sister's children to play with again, and for them to make happy memories too.

So expert metalworker Dom Chinea steps forward to take a look. The paint has worn off and the metalwork of the train is extremely rusty. The pedals don't turn properly. The woodwork is rotten and broken away in places, and there's even woodworm. Yet none of this phases Dom, who can't contain his excitement at the sight of this very special ride-on toy.

The first thing he does is start dismantling the train in the outdoor workshop. He sends the metal parts off for sandblasting to remove the rust and dirt. Some of the plywood parts crumble away as he removes them, but he

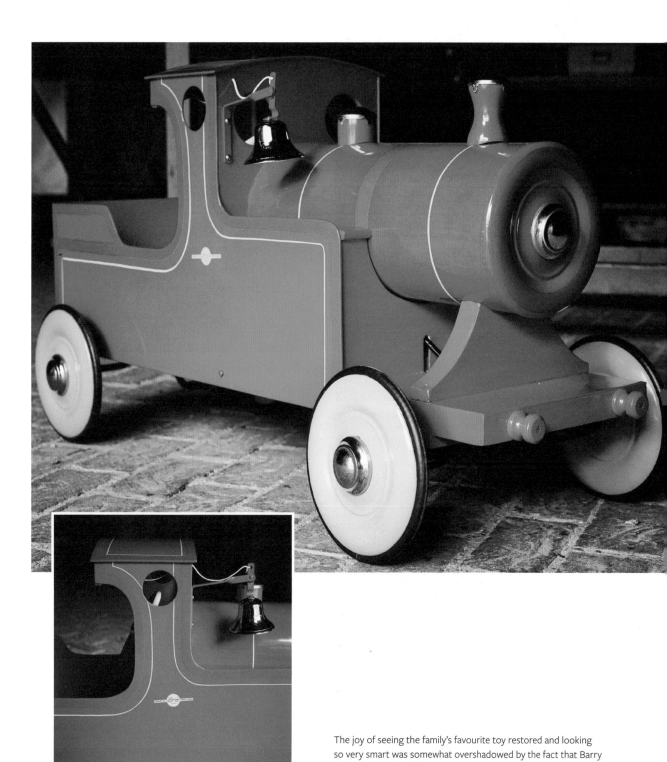

The joy of seeing the family's favourite toy restored and looking so very smart was somewhat overshadowed by the fact that Barry passed away before Dom had completed his work.

can use the shapes to make templates for birch plywood replacements.

Slightly worried that new and old won't fit together well, he clamps everything together first to make sure it all lines up correctly before gluing into place. The train's chassis, he explains, is almost like the foundations of a building; it has to be straight and level for everything to be built up properly. Thankfully, new and old line up perfectly!

There are mixed results from the sandblasting, however. The pedal mechanism and the steering are fine, but the wheels have a few holes, which Dom must weld to make solid again.

Finally, Dom sets to work repainting all the rejuvenated metalwork and plywood bright red, picking out the green and gold details.

Steaming ahead now, it's time to add the wheels, making sure they're greased up properly so they spin freely. The toy train is transformed from the wreck Kate brought in, Dom says.

When Kate returns with her daughter, she brings sad news. Barry has now passed away, but she says with a brave smile that the train will be a nice way to remember him. Agnes rings the train's bell and everyone is overjoyed as Kate pushes her around the workshop.

Repainting the pedal train is the final step – and Dom's favourite job – in giving the family toy a new lease of life.

Rocking Duck

Restoring a family favourite

Some of the items brought into The Repair Shop carry the weight of history with them. Others just make us smile. This was definitely the case with a whimsical 1950s rocking duck chair belonging to Nick Cope from Doncaster, who wants his one-year-old grandson, Fred, to enjoy the fun family heirloom just as much as he did. Nick, a physiotherapist, says that it was bought in 1959 by his grandma for his aunt, Michelle, who tragically died in a road accident when she was just four years old.

Although 'Michelle's duck' has now been enjoyed by three generations of Nick's family, it's especially precious to his mum, Margaret. She was 20 when her little sister Michelle was born and they spent a lot of time together.

Nick says it would mean a great deal to have the ride-on toy restored for his grandson – Fred is his daughter Molly's son – to use and for his mum to see her great-grandson playing on it, just as Nick and his two brothers, his own children and his nephews and nieces did . He feels that if it was restored to tip-top condition, there's no reason why it can't keep on rocking in the family for further generations.

Over the years Nick's mum has spoken a lot about Michelle, sharing stories with Nick and his brothers; this heirloom is a poignant reminder of Michelle's short life. Nick's elder brother, Simon, was born a year after Michelle passed away. Nick says the rocking duck was last used in about 1997. He would like to see it functional again as his grandson is now just the right age, and it would please his mum. He thinks she would be 'overwhelmed' to see it restored.

The poor old thing has certainly seen better days. Its wooden head and body are wobbly and worn, the paintwork chipped and flaking. The seat upholstery is battered, and the metal safety bar went missing years ago.

The man for this job is metalworker Dom Chinea. His plan is to touch up the paint work in the worst areas but maintain as much of the original paintwork as possible so that the duck can keep its soft lemon and duck egg blue pastel vintage good looks.

To make playing with it safe again, he needs to research what the missing safety bar looked like and then take the time to fashion a new one. Nick's mum says that when he was small, he would rock on the duck so hard he could propel it from one side of the room to the other. The bolts need tightening up too, to stop the duck's head from wobbling in a slightly alarming manner. Sonnaz Nooranvary pitches in to make new upholstery and Dom attaches this to the metal frame.

When he returns for the reveal with baby Fred in his arms, Nick can't stop beaming. Their beloved family rocking duck is now like new and ready to go quackers with the next generation.

'Michelle's duck', with its family history preserved in Dom's sensitive restoration, is once again ready to ride.

the
teddy bear ladies

Julie Tatchell and Amanda Middleditch

Some of the most heartwarming moments on *The Repair Shop* occur when an owner is reunited with a beloved childhood teddy bear or cuddly toy, repaired and restored to something approaching its former lovableness. The duo behind these magical transformations are Julie Tatchell and Amanda Middleditch, affectionately nicknamed 'The Teddy Bear Ladies'.

The pair first set up business together in 2006. At that time, Julie had a shop selling high-end handmade teddy bears, with a craft studio attached. She asked Amanda, who was already making her own bears and selling them through the shop, to work in the craft studio and demonstrate her repair skills. Rather than line the repaired toys up willy-nilly on a shelf, Julie and Amanda would construct little tableaux – such as a bear hotel in an apple crate, as if the toys had taken a much-deserved holiday. As time went on, they realized that there was much more interest in the repair side of the business than in the new bears, which is when the partnership started to move in that direction.

Identifying the origins of the historic bears they are asked to repair is a key element of their restoration work.

At first there were hurdles to overcome: advertising their skills only took them so far, and clients outside their immediate local area were reluctant to leave their precious bears with people they hadn't met, or to entrust them to the post. Undaunted, Julie and Amanda set up the first ever touring pop-up teddy bear clinics, visiting about five or six different locations throughout the year. At these pop-up clinics, people had the chance to meet them and they could bring in their toys to be assessed. The pair would then take the toys away for repair, returning in about six months to hand back the restored items.

It was during this period that Julie and Amanda began their association with the Merrythought company, the only surviving British teddy bear manufacturer. The Merrythought shop at Ironbridge, one of the locations for their pop-up clinics, was the furthest they ventured from their home base in the south.

Merrythought – the name is a local Midlands word for 'wishbone', which is the company trademark – was founded in 1930 by the great-grandparents of the present owners, is still located on the same site, and continues to be owned by the same family, now staffed by the third generation of workers. Whereas once the factory sold a wide variety of push-along toys, pyjama cases, and a huge range of bears, today they specialize in the collectible market, making high-end bears for upmarket retailers such as Harrods. They have a big overseas following and are especially popular in Japan.

Over the years, The Teddy Bear Ladies have built up a wide knowledge of other teddy bear brands and manufacturers, from traditional British companies such as Chiltern to the world-famous German Steiff brand. Identifying the origins of the historic bears they are asked to repair is a key element of their restoration work.

Eventually, Julie and Amanda arrived at the point where they were starting to become better known and had so much repair work to do that there was no need for them to keep travelling around the country with their pop-up clinics. Today, on *The Repair Shop*, the skills and sensitivity they bring to each restoration can be appreciated by a wider audience than ever before.

Goldie Bear

A teddy gets his growl back

Nerys Kendrick brings in Goldie Bear to The Repair Shop in the hope that this much-loved teddy can be restored to his former handsome self. Passed down three generations of the same family, from Judith Evans to her daughter Nerys, and then to Aaliyah, her 10-year-old granddaughter, he's been a best friend and confidant to little girls for more than 70 years.

However, poor Goldie, whom Judith thinks was manufactured by Merrythought, a traditional British teddy bear manufacturer, has certainly been through the wars. Well-intentioned amateurs running repairs have left the once-proud bear – who arrived one Christmas morning with a big red satin bow around his neck, to the delight of a very young Judith – looking rather sorry for himself. He's had his nose sewn back on upside down and his mouth restitched under his chin. Faux fur has replaced the original velvet. His gold fur – hence the name Goldie – is now balding in some areas, and he could also do with a bit of extra stuffing to plump him up. He's even lost his growl.

Four individual children, including Nerys's sister, Kay, have adored Goldie Bear. Nerys points out that he will be almost a century old by the time Aaliyah has her own children and he comes to be loved by the next generation.

Nerys would also love dear old Goldie to look his best for his future responsibilities. He has been much more than a toy for the three generations – not only a confidant, secret-keeper, best friend and brother but also honoured teddy bear's tea party guest, where he always had the best seat, the best cup and saucer and an extra slice of cake.

The need to give Goldie an urgent makeover has been given added impetus because Judith suffers from a condition that could lead to her going blind. It's now even more important for her to see Goldie now and as she remembers him.

Bear doctors Julie Tatchell and Amanda Middleditch treat this responsibility with grave attention. They will have to pull off a careful balancing act – Nerys has respectfully asked for Goldie not be transformed into an entirely new bear, and they agree that Goldie must continue to be Goldie but with a new lease of life.

To work as effectively as possible, Amanda dismantles Goldie, so they can operate on him piece by piece. She carefully removes the growler from his tummy and agrees with Julie that it needs repairing. When Judith was a little girl, she would ask Goldie if he wanted sugar in his tea. She would then tip him to the right and to the left so that he growled twice for 'yes'.

Musical instrument restorer David Burville is the ideal person for this job, and he explains that the growler is just a really simplified version of a harmonium or reed organ. He removes the end cap from the mechanism and takes out the bellows part. He can see that the cloth has become

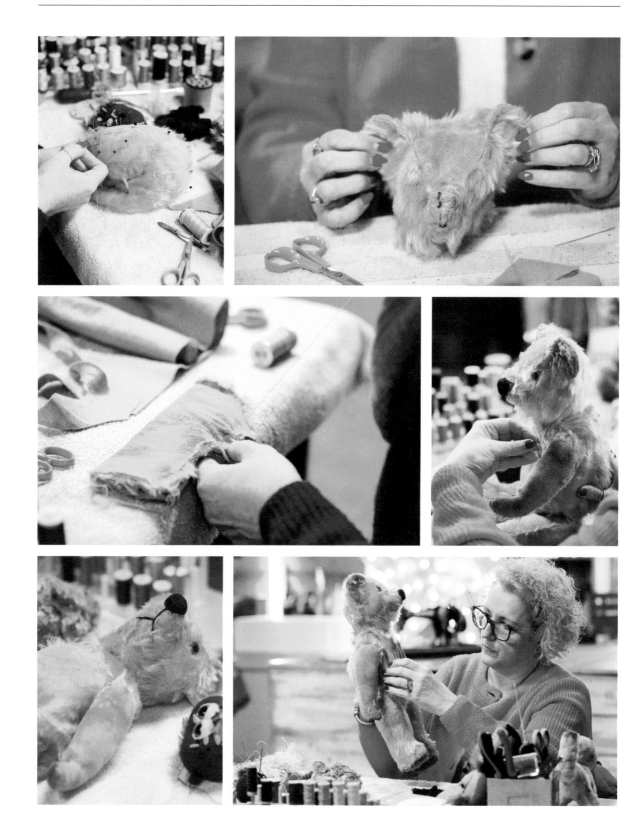

unglued, which he believes is the main reason why the growler doesn't work. So he sticks the cloth back down, then gives the cloth a coat of rubberized paint, to make the bellows part completely airtight and able to function.

Amanda explains that bears of this era are typically missing their noses. To remould Goldie's, she decides to use a plastic nose as a mould and cover it in black felt, making indentations for the nostrils. She saturates the felt in PVA glue, which dries clear and hard, and covers the form with it. Finally, she stitches Goldie's new nose in place.

Meanwhile, Julie strengthens Goldie by giving him some new stuffing, and she covers the thinning faux fur on his paws with cotton velvet, to match the original material.

After leading such a busy life, it's no surprise that Goldie needs a good clean, in particular removing a stain from his ear. Julie and Amanda's work requires painstaking care and, amazingly, it unearths a whole new

identity for Goldie. They suspect that rather than being a Merrythought bear, he is, in fact, an equally prestigious Pedigree bear.

When he has been gently sewn back together, with his repaired growler safely anchored in his chest, Goldie looks like a different bear – but not too different. As a finishing touch, Amanda makes him a yellow raincoat and a hat, just like the outfit Nerys's grandmother had made for him out of Judith's 1960s PVC coat.

There are more than a few tears when The Teddy Bear Ladies hand Goldie back to Judith, Nerys and Aaliyah. He is exactly the handsome bear Judith remembers, and everyone is over the moon that he has got his growl back.

Following major teddy bear surgery, Goldie has been given a new lease of life and has returned to befriend a new generation of young children in the family.

Pancake Dog

Every dog has its day

For Sue Gent from Yorkshire, Pancake Dog, a 1960s, battery-operated mechanical toy, brings bittersweet memories flooding back – of playing with her younger brother, Andrew.

This special toy, which was made in Japan, was given to Sue by her parents at Christmas in 1968, when she was six years old. Sue would like to have Pancake Dog working again so she can share the fun with her father, who is suffering from both Alzheimer's and vascular dementia. She hopes that, perhaps, if Pancake Dog is mended, he will spark recollections of those happy childhood days in her dad, so his face will light up once again and they can laugh together. Sadly, Sue recently lost her mum and regrets not getting Pancake Dog fixed in time for her to see it.

Pancake Dog was Sue's favourite toy, and the only one she has kept from her childhood, because he reminds her so much of Andrew, who died a year after being involved in a car crash when he was just 25. Andrew was four years younger than Sue – but at more than six foot tall, much bigger, she says with a smile. They were very close, however, and when they were children, they loved playing together with Pancake Dog, giggling as he jumped up and down and shook his pan.

As the children grew up, Sue's parents kept Pancake Dog safe, but presented him to her when she married and left home. Sue is not sure when or why Pancake Dog

stopped working, but she knows she could never part with him. She has always intended to have him restored, and admits it's become a family joke that she's going to take Pancake Dog to a mender's and see if he can be fixed. That's why she has brought him to The Repair Shop.

When he's in full working order, Pancake Dog's eyes and ears pop up, his stove lights up red and his pan moves to flip the burgers – he's never flipped a pancake in his life, though. He 'cooks' burgers and his official name was 'Burger Chef'.

To Sue, his name doesn't matter; he's her Pancake Dog and always will be. And now she would love nothing more than to see him working again. She's looking forward to the practical and creative collaboration between mechanical mastermind Steve Fletcher and soft toy experts Julie Tatchell and Amanda Middleditch.

Pancake Dog's wires and connections have corroded so that he no longer lights up or moves. His stove has lost power, too, and his cute chef's outfit – jacket, trousers and apron – are worn and falling apart.

Steve begins by dismantling the toy and removing the mechanism. Next, he tackles the rust in the battery compartment, checks that the motor and wiring are working, and makes the necessary delicate repairs to get the mechanism going again.

To fix the eye and ear movement, he needs a steady

hand to straighten out the metal and prepare the levers before Julie and Amanda can take over. Meanwhile, the ladies have removed the dog's tatty clothes and have made templates to make new clothes to match the original. The only piece they are able to save is the apron. Finally, they attach new fur to his head and ears, and give his eyes and accessories a thorough clean.

Sue is overjoyed when she witnesses this amazing revival, finding it difficult to believe that Pancake Dog now looks exactly the same as he did when she was a child. She can't wait to take him home to see her dad.

The combined skills of The Teddy Bear Ladies and mechanical expert Steve Fletcher have transformed Pancake Dog back into a smart and efficient short-order cook.

the
woodworker

Will Kirk

If his mother hadn't stumbled across an advertisement at the back of a newspaper for a course in antique restoration at London Metropolitan University, Will Kirk might never have discovered his passion for furniture restoration and woodworking. Up until then, he had never dreamed that such a course existed.

Will was always much more drawn to creative subjects than academic ones. After he finished secondary school, where the discipline and boundaries were quite strict, he spent a year at Camberwell School of Art, which really opened his eyes. Here he was given the freedom to express himself and explore different creative avenues.

From Camberwell he went on to study graphic design at London College of Communication. Yet, although the creative side of the course appealed to him, he had the sense that he was taking himself down a route that was ultimately not going to be entirely fulfilling – part of the reason was that the work was so computer-based and not very hands-on.

It was around this time that his mother showed the ad to him. What followed was a real turning point. As soon as Will embarked on London Met's three-year BA course in antique restoration and conservation, he knew he had finally found his direction in life. From there he never looked back. Funnily enough, it was not until he had started the course that he remembered his grandfather had also enjoyed restoring antique furniture, so there was already a precedent in the family.

After his degree, Will went to work for several major British furniture restorers, before setting up his own workshop in Battersea. Here he gradually established a successful business. Through trial and error, he added new techniques that he picked up on the job to the skills he had been taught at university.

Over the years he has worked on precious family heirlooms and corporate commissions – restorations involving everything from gilding and cabinetmaking to carving and French polishing. His expertise is such that he has travelled to Italy with The Worshipful Company of Painter-Stainers to represent Britain at the annual Salon Decorative Arts Fair.

One of Will's clients, for whom he was restoring a chair, worked for the production company that would go on to make *The Repair Shop*. He asked Will if he'd ever thought about appearing on television. Despite having already taken part in a couple of shows, Will wasn't entirely sold on the idea. Six months later, when *The Repair Shop* was in development, the production company was back filming Will in his workshop.

As soon as *The Repair Shop* was given the go-ahead, Will was asked to join the show. From then onward, he has been one of the key experts in the barn, relishing the challenges that come through the door.

A key question in restoration is how far do you pull it back? How much of the past do you leave in? If something is structurally damaged, you have to get it back to working order. But aesthetically it's always nice to leave those nicks and scrapes because that gives the furniture its age and character.

We are always led by what the owners want us to do. They might have scribbled their name inside a box when they were a child and now, years later, they want to keep that mark as a reminder. But there's also an ethical dimension. If someone brought me a lovely Chippendale chair and asked me to paint it black, I'd say no.

Many of the items that come to me are in boxes of bits, so the main task is piecing them all back together again. Heat and water damage is also common.

One of my biggest challenges was an octagonal table inlaid with very small pieces of wood and mother-of-pearl. It came in absolutely knackered, a prime example of something that looked like it should have been put in a skip or on a bonfire. Someone had set a hot pan down on the surface and a lot of the inlay was missing. I had to remake everything that was missing, including the central portion, which included an Arabic inscription. Although fiddly jobs like these take hours, I really enjoy being able to save something from the brink. When you see the craft that has gone into the original piece, it's so satisfying to rescue it and stop it from going to landfill.

People often don't appreciate how much damage light can do to wood. They might place a lovely rosewood table in front of a bay window and put a fruit bowl on top, which looks great when the neighbours are walking past, and only realize how much the sun has bleached the wood when they come to move the fruit bowl. Either you can colour the bleached area to match the darker patch, or bleach the entire top, and recolour and polish.

In the case of a burn, I'll remove a lozenge of wood where the mark is and stick a new piece of veneer in its place to match. It's not just a question of replacing the same type of wood, mahogany with mahogany, for example. You have to source and match the grain and the colour. I have drawers and drawers of veneer in my workshop – nothing ever goes to waste.

Scratches and cracks I repair with wood filler, which is sawdust mixed with glue. In my workshop, I normally use a traditional restorer's glue. If I'm filling a crack in an oak table, for example, I'll mix oak sawdust with glue. In the barn, the glue might be PVA, which dries faster. Otherwise, shellac sticks are really great. I always have some in my toolbox – they're the furniture restorer's best friend. The sticks are a mixture of wax and shellac and come in billions of shades of brown, so you can

match the colour of the surface. You melt the stick into any chips and scratches in the wood and chisel off the excess with a sharp edge.

One thing I learned at university is the importance of preparation. To achieve a lovely polished surface, you have to take your time. First, you have to remove the old polish – on an old piece this is most likely made of shellac, which is soluble in alcohol. I use methylated spirits and either a clear cloth or ultra-fine steel wool. Then it's a question of sanding down the surface, working your way through all the grades of paper, from coarsest to finest.

French polishing is a technique, not a type of liquid. Most antique furniture is polished with shellac, which is brushed on. French polishing, which gives a shiny mirrored sheen, involves diluting the shellac with methylated spirits and applying it in thin coats with a pad. It takes a lot of patience and practice. So does mixing pigments with the polish to colour-match the wood grain. I've been doing it for so long now, it's second nature.

Otherwise, it depends on the piece. Mid-century furniture, for example, tends to be finished in teak oil, Danish oil or linseed.

The furniture restorer's worst nightmare is woodworm. The woodworm beetle lays its eggs in the grain and grooves of the wood, then the larvae eat into the wood and metamorphize into beetles, which in turn work their way back out and fly away. The characteristic tiny holes are actually flight holes.

If there are tiny holes in the surface of the wood, that's a strong indication of woodworm. But sometimes they're not immediately apparent. Softer, sappier wood, such as beech, is more susceptible – the pests prefer timber with a high moisture content. An antique chair may have its back and legs made of more expensive wood such as mahogany or rosewood, but the seat frame itself will be a cheaper timber because it isn't designed to be seen. Once you take off the upholstery, that's when you can see the damage.

The first thing you need to do is work out if the woodworm is active or not. Hold a sheet of white paper under the holes and tap on the wood. If what comes out is grey powder, that suggests an old infestation. If it looks more like sawdust, there's a good chance the infestation is active.

'One thing I learned at university is the importance of preparation. To achieve a lovely polished surface, you have to take your time.'

Furniture that has woodworm sounds hollow when tapped. Inside it will be very brittle, like a honeycomb. There are solutions you can apply to the surface to harden it, but essentially you're only hardening the honeycomb, you aren't making it more robust.

If the infestation is active, you can inject the holes with insecticide, but it will take hours and hours and it won't necessarily work. Often the most cost-effective way to treat it is to send the piece off to a specialist company. The piece will be put into a chamber and the heat turned up very high, which will destroy any infestation without damaging the wood itself.

One of my favourite tools is a big chisel I bought when I was in the market in Bath with my father-in-law. It's about a hundred years old and was made in Sheffield. You can almost use it like a plane because it's so wide and sharp. Sheffield steel really keeps its edge. It feels right in the hand and you have that weight behind you when you are shaping things up. I always buy old tools if I can. They're cheaper than new, better quality and last longer.

People know I enjoy carving, so I have at least a hundred carving gouges, many of which have been given to me. I also have a lovely mallet, which is made of olive wood and which cost a fortune. When you use it, it smells of olive oil.

One of the things I like most about working on *The Repair Shop* is that everyone shares and shares alike. Some old furniture restorers won't give away their trade secrets – they won't pass on their special recipe for furniture wax or whatever, which strikes me as bizarre. Why run the risk of losing that knowledge? Here we all have massive respect for one another and we're all willing to learn from each other. As Jay likes to say, there's over 600 years of experience in the barn. I think you could bring anything in and we'd be able to fix it.

Turkish Bath Clogs

Putting the best foot forward

Occasionally on *The Repair Shop*, the most unlikely item can turn out to speak volumes – of war, revolution and exile. This is particularly true of these delicate hammam bath clogs from Turkey, known as *nilan* in Turkish. They were brought in by 90-year-old Peter Shahbenderian, a proud descendant of a well-to-do Armenian family of traders who left Turkey for Paris because of the Armenian genocide during World War I.

Any property and possessions left behind were confiscated by advancing troops, so whatever the family managed to keep with them came with added provenance. Peter's grandfather started life again in the French capital, setting up a textiles shop on rue de Rivoli.

In 1925, at the age of 18, his daughter – Peter's mother, Zarmine – left Paris and moved to Manchester to marry

Gregory, the son of a manufacturer who was part of the thriving Armenian community there. Armenians had settled in this northern industrial powerhouse in 1835 as silk merchants and became very influential in the textile industry.

Zarmine brought her *nilan* with her, as part of her trousseau. Peter recalls seeing this collection of special clothes, hats and shoes in the attic, but everything, apart from the clogs, has long since disappeared. They are all that Peter now has of his mother's personal possessions.

Peter, a historian and Armenian expert, explains that, as a girl, Zarmine would have worn her clogs whenever she visited a Turkish bath, or hammam, in Constantinople, now Istanbul. Inlaid with delicate mother-of-pearl, with leather and copper wire straps, the clogs were probably made at the end of the 19th or early 20th century.

Zarmine kept this remarkable reminder of her early life on a high shelf in the bathroom at home. When she died in 2005, aged nearly 100, the clogs passed to Peter, who lives in London. He says that they must have been treasured by his mother for her to have kept them all those years, but she rarely spoke of them. He thinks that perhaps there was always an element of her not wanting to dwell on the sadness of the past.

He considered restoring them some years ago, but jokes that at his advanced age he no longer has the ability

to carry out such detailed, painstaking work.

Peter's second wife, art historian Clare Ford-Wille, whom he married after the death of his first wife, Nuné, feels lucky to have had the chance to get to know her mother-in-law, describing her as a calm and lovely person.

Damage has occurred to Zarmine's clogs over time. Many pieces of mother-of-pearl are missing and the straps are worn. The fine copper wire on the straps is badly tarnished and coming away from the leather, which is cracked and falling apart.

Wood expert Will Kirk starts the ball rolling by carefully removing the perishing leather straps from the clogs and handing them over to leather expert Suzie Fletcher so she can treat and repair them. She also buffs the copper wire on the straps with fine steel wool, giving a lustre to their dull surface.

Taking a deep breath, Will moves onto the most challenging part of the job. Using sustainable mother-of-pearl, sourced to match the original inlay, he embarks on a complex operation, hand-cutting, shaping, filing and gluing 80 to 90 tiny lustrous fragments to replace the missing parts of the intricate marquetry. He has to be precise, quick and sure in his placement of every single piece.

Finally, Will selects a sheet of finest conservation grade sandpaper and very gently sands and cleans the clogs. He uses a Danish oil to seal and nourish the surface, as well as providing a better water-resistant surface just in case the clogs are kept in a damp environment again. He then hands the clogs to Suzie, to reattach the leather straps. She does this with regular shoe tags, which she covers with a piece of leather.

At the reveal, Will and Suzie hand the priceless clogs back to Peter and Clare, who are absolutely lost for words at their transformation. Eventually, Peter will pass the clogs on to his own daughter, Mané, as a family heirloom, but before that he intends to honour his mother's memory in a way she would have approved of: he will take the clogs home with him and display them as the treasured objects they were for her and still are for him.

Intricate, painstaking work has breathed new life into these precious Turkish bath clogs inlaid with mother-of-pearl, which were part of Peter's mother's wedding trousseau.

Rickety Table

Making new memories

The plight of 12-year-old Jacob would crack the hardest of hearts, and The Repair Shop was privileged to step in and rescue the rickety pine gate-leg table that keeps the memories of his beloved daddy, Barney, alive.

Barney died suddenly in 2015, when Jacob was only six years old. Barney had suffered from neurological and physical problems after being involved in a near-fatal car accident in 2010. As a result, Jacob moved in with his grandma, Heather.

Every Sunday, however, Barney would join Jacob and Heather for a delicious home-cooked Sunday dinner and to take part in fun arts and crafts activities around the table, which Heather remembers from her own childhood. She explains that these Sunday dinners were the highlight of Jacob's week, as he so enjoyed the quality time he spent with his father.

After the car accident, Barney was in intensive care for a week. If he hadn't pulled through, Jacob would have no memories of his daddy at all. He and Heather are so grateful for the precious time they all spent around this table. It has helped to give Jacob confidence to grow up into a bright and articulate youngster.

The table has been in the family for generations and Heather has fond memories of her children at the table, recalling how Barney would sit and draw round the knots in the pine with a pen.

The table is now past its best. It's functional in that the gate legs still work but some of the joints are loose, so the legs are rickety and the whole table wobbles. There is a large crack down one of the centre planks, and the surface of the table is extremely worn, where the pine has been rubbed down over the years with a damp cloth. At some point – and Heather struggles to remember when – it was varnished, but there is very little of the varnish left now. She keeps threatening to get rid of the table and replace it with a new one, but Jacob knows she is only joking.

There is so much shared family history in every joint and nook and cranny of this table. Jacob and Heather tell antique furniture restorer and wood expert Will Kirk that when he works his magic on making the table sound and stable again, they would like some of the bumps and grazes kept. They would prefer it not to be completely shiny and look brand new, and they want to continue to make precious memories with it.

Even for an expert like Will, the job turns out to be a bit of a challenge. He explains that it would be far easier to simply replace the very dry and warped top panels with new pine but that isn't an option because he knows how much the complete table means to Jacob.

After disassembling the tabletop, Will sands the old finish off the top of the table. He then soaks a towel in

warm water and wraps it around a piece of wood and steams it for a few seconds with an iron – his aim is to force moisture into the wood, opening up the grain, to make it more malleable so it can be flattened. To his great relief, the process works – the wood is far more flexible than the untreated pieces – and he repeats it for the whole tabletop. He then clamps the pieces to his workbench to help things along. To reassemble the tabletop, Will uses a technique called biscuit-joining, which makes the joints even sturdier.

The rest of the table is in much better condition. To remove the remaining scraps of old varnish, Will uses a homemade furniture-cleaning solution containing methylated spirits, white spirit and linseed oil. He then repolishes the whole table with a water- and heat-resistant shellac polish, much the same as used in French polishing but with a few additional ingredients.

When Will and Jay return the table to Jacob and Heather, they are excited and emotional. Jacob says that he would like to put the table in his own house when he grows up and use it every day with his own children and grandchildren. Will's fix means more than the whole world to this lad and his grandma.

REMOVING A WATER STAIN

'Although water damage can be very tricky to repair, it's not difficult to remove a white ring from polished wood. To do this, you use mayonnaise. Apply the mayonnaise to the white mark and leave overnight, then wipe it off the next morning. The vinegar in the mayonnaise draws the moisture out, while the mayonnaise is thick enough to keep the vinegar in place while it does its job. But you can only do this on a polished surface. If it's dry bare wood, you'll be left with an oily stain.'

Jacob is overjoyed at Will's repair of the table and is looking forward
to having the family sit back around it, keeping the memory of his
dad, Barney, alive.

Wooden Ship

A model ketch sets sail again

When Matt Goddard's father, Stephen, passed away suddenly in 2009, he left behind his devasted 15-year-old son and an unfinished 1/50 scale model of the *Clara May*, which he had started building in 1999. Produced by a highly regarded ship model company, this historic re-creation of an 1891 wooden ketch is approximately 80 years old, still with its original box.

It has taken more than a decade for Matt, who is from Greater Manchester, to come to terms with the loss of his dear dad. But now he would love to see the ship finished. It was the very last thing Stephen was working on, and Matt has never found anyone to take up the mantle and finish the project – until now. He did take a look at the instruction booklet once, but he is afraid that if he attempted the job himself, he would ruin it. He has much more respect for his father's skills now.

Stephen had a workstation in the shed at the bottom of the garden. Matt would sneak in after school and during the holidays to see what his dad was up to and witness the progress he was making with his latest creation. Whenever Matt had a school project to work on, Stephen, who worked in the sign industry, would magic up the most amazing items for him.

Once they found an old kayak in a relative's garden, which was going to be thrown away. But Stephen saw an opportunity to rescue the neglected vessel and, over the course of six or seven months, it became a father-and-son project. Together they patched it up. Stephen even created graphics to make it look like a racing kayak and they took it out together on the River Dee, near their home.

However, building the 50cm-long model ketch was Stephen's true passion project; he took his time over it, considering every aspect in painstaking detail. Matt has fond memories of seeing the ship when his dad was working on it. He always thought it looked quite spectacular but, to his disappointment, he was never allowed to play with it.

When Stephen passed away, he left few possessions, so the unfinished ship has been particularly treasured by his son. It is between 80 and 85 per cent complete, but the remaining tasks are fiddly and require the skill and

patience of wood expert Will Kirk, who is keen to start work and finish this very special father-and-son keepsake.

The structure is complete but most of the sails and masts need reconstructing and attaching. The sails that Stephen did perfect have rusted slightly where the fabric has been nailed to the mast. Matt explains that the ship has also gone through several house moves and been rather bashed about in transit.

The top of the forward deck is chipped and there is scuffing on the bow. Having been left unattended for so long, the wood is now dry and discoloured. After assessing the damage, Will removes the glue from the broken sections, cleans the whole thing thoroughly and reattaches the wooden pieces for the main body of the ship.

Teddy Bear Lady Amanda Middleditch is consulted for her expert advice on how to finish the tricky and delicate sails. Once these are completed, the masts are attached to complete the model. And then, before the reveal of the finished ship to Matt, Will's final task is to create a smart display stand.

Matt says that his father was fantastic, a truly wonderful man. For him and his family, the restored *Clara May* is an amazing tribute to his memory and one that they will be able to look at and appreciate every day.

The now shipshape Clara May will deservedly take pride of place in Matt's home in memory of his father.

the leatherworker

Suzie Fletcher

Suzie Fletcher's passion for leatherworking arose directly from her childhood love of horses. An outdoorsy child, who started riding at a very early age, she was fascinated by the paraphernalia of saddles, bridles and harnesses, so much so that she would offer to clean the tack at her local riding stables in Oxfordshire. As well as the connection to horses, what attracted her to saddlery so strongly was its beautiful combination of form and function. She always enjoyed making things, an artistic side encouraged by her parents, and soon she was cutting up scraps of leather to see what she could fashion them into.

When Suzie came to leave school at age 16, master saddler Ken Langford, for whom she had been working during school holidays and at weekends, suggested that she apply for the saddlery course at Cordwainers College in London. She did and was accepted. A condition of her place was that she complete a year-long pre-entry introductory course, where she studied basic techniques in making footwear and other leather goods – such as handling tools and hides – skills that were to stand her in good stead decades later when she joined the team of experts at The Repair Shop.

At the time, however, her principal love and focus remained saddlery. After college, and a series of interim jobs, she finally obtained a coveted placement with Ken Langford, before going on to study harness-making under Bill Turner in Pangbourne. Then followed a stint with Geoff Dean in West Sussex, learning about bespoke saddlery. Dean, who has since retired, once counted Prince Philip among his clients.

Now with a thorough grounding in saddlery, Suzie's outward-bound spirit next took her to Boulder, Colorado, where she worked in a tack shop. Three weeks after her arrival she met the man who would become her husband. Together, for the next 20 years or so, they lived and worked on a homestead on the eastern plains, tending their animals and managing their small dry acreage. While Suzie built up her reputation in the States as a bespoke master saddler, she was also the authorized US repairer for a number of saddle-makers based in Walsall in the UK.

Then tragedy struck. Her husband was diagnosed with terminal cancer and she devoted herself to making his remaining time as rich and fulfilling as possible. Within five months of his death, she had also lost her mother and her aunt. A few years later, she sold the homestead and returned to the UK in something of an emotional tailspin, with nowhere to live and no workshop.

Step forward The Repair Shop. Her clock-restorer brother Steve Fletcher (see page 82) was already working as an expert on the show, and she'd jokingly said to him that she was available if they ever needed a leatherworker, never expecting anything to come of it. But before she had even left the States, she was contacted and asked to be part of the team, which is how she ended up in the barn a mere five weeks after her return to Britain. Over the years, several saddles have come in for her expert repair; more often, it's all manner of treasured leather items, from old satchels to shoes, that require her restorative touch.

A lot of the leather items that come into The Repair Shop are dirty, dried up and cracked. My job has necessarily evolved from saddle-repairer, designer and maker to learning how to preserve very, very fragile leathers. I've sought out the help of many wonderful conservators, leather chemists and other supportive experts who have been willing to offer their experience and advice. The Repair Shop has allowed me to give a new lease of life to severely damaged, but well-loved, leather items that would normally be too time-consuming and expensive to repair.

Looking after leather is vitally important. All hides contain natural oils or greases. When leather becomes dry, the fibres shrink and will eventually snap. Basic cleaning with saddle soap, which is very mild and contains glycerine, helps to open up the pores. Then you can put on a cream conditioner, which will hydrate the fibres without saturating them. This keeps the leather malleable and stops the fibres from breaking.

When cleaning leather, it's essential to know what kind you are working on, how it's been tanned and whether or not it has a false coating, which won't accept any feeding or conditioning. Hide for saddlery mostly comes from bovine sources, which provides the right degree of thickness. Pig-hide, which is thin and super strong, will take a lot of wear, which means it can be used on the seat of the saddle. Furniture and fashion leathers, on the other hand, tend to have gone through a splitting machine, which means they are fibrous on both sides. Those vibrant blues and pinks and greens are the result of a false finish applied to the front. While the underside might be leather, you can't feed the surface at all.

If there are small gaps, holes, scuffs, scratches or gouges in the surface of the leather, you can fill these with leather filler. Leather filler, which is similar to wood filler, consists of sawdust leather mixed with glue. Once dry, you can stain it to match.

I like to work with natural materials as much as possible. This means I stick to cotton linen threads, not synthetic ones. I'm very aware of sustainability in today's world. So long as it is looked after properly, leather can last for hundreds of years in one piece. Man-made products degrade with use, but are not biodegradable. Leather is. It won't leave a blot on the landscape.

'None of us here "works" for a living.
What we do is who we are. It's a
wonderful way to spend your life.'

I love tools and I have a huge selection that I've collected over the years, everything from knives, needles and awls to pricking irons and crew punches. Some of these have been made specially for me. I also have several lovely sewing machines. There's "The Beast", a German-made Adler, for stitching through thick leather; "Edith", another Adler, for medium-weight work; and "Pearl", a Singer treadle machine with a revolving foot that allows me to sew in any direction.

But my absolute favourite tool is a pair of ordinary pliers that used to belong to my grandad. After college, when I was waiting for a placement with Ken Langford to become available, I set up a little workshop in my grandparents' house and kept my hand in making belts and bags for friends. The room where I worked was my dad's old bedroom, and these pliers just happened to be lying around. There was no formal handover, I just used them and kept them. Today I wouldn't be without them. Later, when my father passed away and we came to sell the family home, Steve asked me if there was anything I'd like from my dad's workshop, which had remained untouched for years, his tools still laid out on the bench. I chose some of my dad's finer pliers. In turn, when it is time for me to go, I plan to gift all my tools to The Worshipful Company of Saddlers, so they can distribute them to worthy craftspeople and students. Tools need to be passed on and used.

So does knowledge. No one person owns all the information, so it is important to share and be reciprocal, to keep on learning, to work out where improvements can be made. I've learned so much from so many other inspirational leatherworkers who have honed their skills beyond the exceptional. Here at The Repair Shop we have this amazing platform, which means viewers can see that it is possible to turn a passion into a career. None of us here "works" for a living. What we do is who we are. It's a wonderful way to spend your life.

Shakespeare's Works

Preserving a miniature library

Isla Smith's granny, Kathleen, was only 12 when she won a school competition for her needlework. Although she wasn't an avid reader, she chose a miniature library of the complete works of Shakespeare as her prize. This now belongs to her granddaughter, Isla, who would love to see the leather-bound case and its contents restored to their original condition so they can be handed down to future generations of her family.

These miniature libraries were very popular in Victorian times, when innovations in printing enabled books to be made with the tiniest type people had ever seen. Publishing houses would compete with each other by creating miniature versions of popular books, in an attempt to show off who had the most accurate and precise printing technology. Shakespeare became the most popular author printed in miniature form.

In Isla's tiny inherited library, his works sit in a neat leather case with a metal clasp and leather hinge, so the top can be opened like a literary treasure chest. However, this precious case is battered, the gold letters of 'Shakespeare' almost worn away, and the condition of the collection so fragile Isla fears it will be impossible for her to leave it to her own children and grandchildren in turn.

The leather on the top side of the case, ripped and bleached by exposure to the sun, needs mending and re-padding. Half of the metal clasp that locks the case is

missing, and the leather hinge has come adrift. And the books themselves – Isla says that, being Scottish, *Macbeth* is her favourite – are showing slight signs of age on their leather covers.

Isla explains that the collection became granny's pride and joy because it represented her achievement in needlework from an early age. Kathleen, awarded her library prize in 1897, went on to become an extraordinarily talented needlewoman, embroiderer and calligraphist, teaching in numerous colleges across the UK.

The Shakespeare library in its leather case passed to Isla after her grandmother's death in 1978, who welcomed it into her possession.

Her granny, whom she describes as 'a strong cup of tea', was strict and old-school, but she encouraged her to not only embrace reading, but to be self-confident and creative. She would often give a piece of fabric to Isla and simply say, 'Make something.'

Spending time with her granny was an important part of Isla's childhood. She learned to cherish the precious Shakespeare library and her granny's memories of the grandfather she never met. He died three years after he and Kathleen were married.

Leather expert Suzie Fletcher starts by removing the top cover of the case and replacing the padding. Taking great care, she then traces a line along the case's cover to

create a template of where the leather was missing, cutting new pieces to replace the missing parts perfectly.

The case is so worn that it's difficult for Suzie to tell what the colour of the leather should be. After a thorough treatment with a conservation grade leather dressing, its original beauty shines through, and she dyes the new leather to match.

Suzie fills in any gaps with flexible leather paste. This is intricate work. She fixes the top cover, without the repaired leather element, to the back of the box, and treats the rest of the leather, fixing anything that's loose. Then she reattaches the repaired leather neatly onto the top of the case and, finally, buffs the leather book covers with a lint-free cloth to bring back their lustre.

Isla says that her granny would have burst into tears if she had been in The Repair Shop to see her Shakespeare collection restored, even though she had never been the type to let her emotions get the better of her. Kathleen's legacy is now in the best of condition for Isla to keep the

family tradition going, passing the miniature library on to her own children and, in turn, to her grandchildren and beyond.

The miniature books, owned by Isla's inspirational grandmother, can be enjoyed by her own children and grandchildren now that Suzie has repaired the leather box and covers.

Canadian Saddle

Sprucing up a family saddle

When, at the age of 18, Elana Beavis finally met her grandmother on her father's side of the family, things she knews about herself started to make sense. Her parents divorced when she was a baby, and Elana never really knew her father's side of the family until she was reunited with them in her late teens. It was only then that she found out her paternal great-grandparents had owned a rodeo farm in the Canadian Rocky Mountains, which she felt explained her lifelong obsession with horses. Sadly, she didn't have much time with her grandmother, but there were enough happy meetings for her to learn something of her family's exciting equestrian history.

Elana turns up at The Repair Shop with a battered – but intricately patterned – heavy leather saddle, which had belonged to her great-grandfather. Eventually, her grandmother inherited it and, when she died, it was passed on to Elana.

Elana says that the saddle is the only thing she has from her father's side of the family and that it feels like a connection to the past and the years she lost with them.

The thrilling and romantic family story reveals that Elana's great-grandparents fell in love in the early 1900s. Her great-grandfather, Morris Averill, was a keen rugby player, turning out for the Worcester and Exeter teams, and even playing against the first All Blacks team to come to the UK.

Morris met Marjorie Pheysey, the daughter of a family of agricultural merchants, but soon after their meeting in 1909 he emigrated to Canada, where he was assigned a ranch plot near Calgary by the Canadian government. Morris and Marjorie didn't forget each other.

Despite the distance and challenging world events, they kept their romance alive by writing countless letters back and forth. During World War I, Marjorie became a nurse, but eventually Morris convinced her to join him in Canada. They married in Calgary soon afterward and had two children, one of whom was Elana's grandmother.

Resident leather expert Suzie Fletcher examines the saddle. She explains that, traditionally, a saddle would be made for each individual rider, rather than for the horse. She believes that the padding on the saddle, numbered 308 and made by Riley & McCormick in Calgary – a famous Canadian saddlery outfitting cowboys since 1901 – was a later addition.

With Elana's agreement, Suzie decides to remove the padding from the saddle and take it back to how it would have appeared originally, with a shearling or sheepskin lining. Typically, Suzie says, English saddles were padded, so this isn't entirely authentic in terms of the Western design. Suzie loves working on this project, as it brings back many happy memories of the years she spent living on a rodeo ranch in the US.

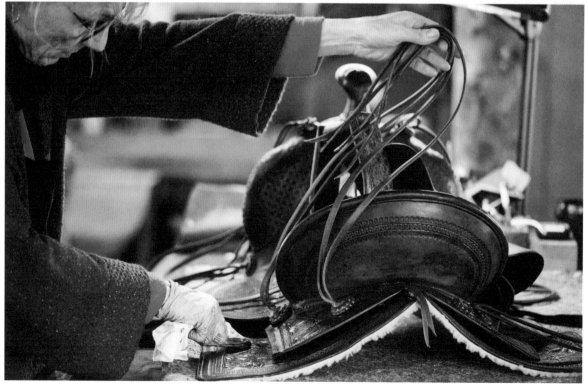

As Elana intends only to display the saddle as a heirloom rather than use it for her own horse riding, Suzie opts to use a faux sheepskin for ethical reasons.

The pretty toothing and tiny Mexican-style flower motifs carved into the leather are still visible, but the saddle has attracted dirt during its travels and the leather has become worn and brittle in places. So Suzie bathes the saddle in warm water to allow the leather to flex and then anoints it with leather treatment oil, buffing up gently to achieve a rich, burnished, chestnut shine.

The leather strings, which would have been used originally used to secure items such as a lasso and blankets, and the circular conchos that hold the saddle together, have perished and need to be replaced. Suzie has discovered that the right kind of supple leather used for Western saddles cannot be sourced in the UK, so asks one of her apprentices in the US to help. She is promptly sent over the perfect grade of leather to do the job and makes new strings and conchos.

Elana is delighted when she sees her very special saddle now restored. She is so pleased that it does the memory of her grandmother's family justice. It is now in good enough condition not only to display at home, but to pass down to her own son, and hopefully in time, it will be be treasured by his children, too.

Elana couldn't feel prouder of her beautifully restored saddle. It is very precious to her and the only possession she has from her father's side of the family.

BESPOKE SADDLERY

'Anyone can make a pretty saddle, but the true challenge is to make a saddle that really functions. Bespoke saddlery means working directly with the horse and rider. Many companies will say that their work is bespoke, even if the saddler does not meet the owner or horse in real life and all they have to go on are dimensions and videos.

'The horse is key. If you think about a horse's back, there are a multitude of muscles that undulate as it performs various movements. The back is not static. Then you put a saddle on top, which is static, and a rider on top of that, who will throw the horse's balance out, and who won't necessarily move with the horse. For the horse not to suffer any discomfort, the saddle has to really fit. It has to fit the rider, too, so there's no tension that might get transferred to the horse.

'A good saddle can assist riding skills, but it won't make you a good rider. I can't make a horse happy if the rider is not prepared to put the work into their own riding ability. I'm just as hard on myself as a rider, and make sure I stay fit so I don't cause any distress to my horse.'

Child's Lost Shoe

Suzie Fletcher and Will Kirk

An evocative wartime talisman

When his father passed away in 1995, Jonathan Sparkes asked his mother if he could have the tiny, leather child's shoe that his grandfather had found on a battlefield in France and had passed on to his father. His mother asked him what he was going to do with 'that tatty old thing', with its wafer-thin leather, missing back and disintegrating lining, but Jonathan simply said that he wanted it because it was special to him.

Although the sole is intact and still has the original nails, the shoe is barely recognizable as footwear. It holds no monetary worth but, for Jonathan, its sentimental value is priceless. It represents a link back to his charismatic father, Ned, and to his grandfather, Hubert, a brave soldier whom he never knew.

Hubert found 'The Shoe', as the family call it, lying in the mud as he marched across France in World War I, and it was to become his protective talisman.

When Hubert picked up the shoe, it looked fairly new and he hoped that he might spot the mother with a child missing a shoe when he arrived at the next village. But the chaos of war churned all around and he never did find its young owner.

Hubert survived the war and so did the shoe. He kept it safe and no one knew of its existence until Ned, his son, was about to set off for paratrooper training in World War II. Hubert pressed this mystery object into his hand and said that he must take it with him; it had kept him safe throughout the fighting, and it would do the same for Ned.

Jonathan was told that his father, a member of the Parachute Regiment, had the shoe with him during the war, but he only found out after Ned's death that he had completed an astounding 126 parachute jumps, many of them behind enemy lines.

Wherever Ned went, the shoe had to go too. It even accompanied him to South Africa when he emigrated in 1953, bringing up his family there. Jonathan, now based in Berkshire, lived in South Africa until recently, before returning to the UK and bringing this well-travelled talisman back with him.

Ned had left school at the age of 14, but achieved the army rank of captain. Through all his adventures, he kept the shoe sandwiched in his wallet in his back pocket.

Jonathan fears that the worn and tattered shoe, which he has kept in a safe, will eventually disintegrate, meaning that a piece of his father's and grandfather's courageous history will be lost forever. He doesn't expect it to end up looking like new, but if it could simply be restored to something his father and grandfather would have recognized, that would be enough.

It would be amazing to see it with a heel and a buckle, he adds, because these would have been in place when his grandfather found the shoe in the mud. But he's happy to give the experts carte blanche to do what they think is best.

So leather specialist Suzie Fletcher and woodworker Will Kirk put their heads together. Will suggests making a wooden 'foot' to go inside the shoe permanently and support the delicate leather upper. But before he does, Suzie dabs consolidating gel all over the existing leather to give it back some structure. Once it's dry, she hands the shoe over to Will with a drawing of how she thinks it should look. He marks out a rough shape of the shoe on a piece of wood, which he cuts with a bandsaw to get rid of most of the excess wood, then follows with a sander and sandpaper to refine the shape. It's a perfect fit.

For Suzie, it's a conundrum as to how to strengthen and line the shoe because the leather is too fragile to be manipulated. She decides to use the technique of wet-moulding, which allows her to stretch the new lining leather around the form so that it dries to the right shape.

After applying glue to it, she pushes it inside the shoe, followed by the form. Once dry, she cuts out the original pattern on the upper through the lining leather with a very sharp scalpel. Finally, she constructs a new leather back, adds a strap, punching a few holes in it, and a buckle.

At the reveal with Suzie and Jay, Jonathan is visibly moved. Having the shoe restored means more to him than anyone could imagine, and he hopes that future generations of his family will treasure it as much as him.

'There are some repairs that seem as though they'll be impossible to do, but the story behind them is so powerful that the experts pool their skills, discuss the options, then collaborate closely with one another to re-create something wonderful.'

Jonathan can hardly hold back his tears when he sees the finished shoe. Suzie, with Will's help, has given him a real, tangible connection to both his father and grandfather.

the
horologist

Steve Fletcher

Third-generation horologist Steve Fletcher wasn't always drawn to the trade, in fact, quite the reverse. Watching his father repair watches all day long, in what seemed a form of self-imposed solitary confinement, and for little financial reward, was enough to convince Steve that that life was not for him. Instead, his love of birds, animals and nature in general made him dream of becoming a vet. During the summer holidays, he accompanied a local vet on farm visits near his home town of Witney, and studied the relevant subjects at school.

Yet academic work proved to be the stumbling block. Despite passing his O levels, school was always quite a struggle for Steve and by the age of 15 he couldn't bear to stay there a second longer. Instead, his creative and inventive bent, which had always gone hand in hand with his love of nature, inspired him to apply for a silversmithing apprenticeship. Although his application was successful, his parents couldn't afford his board and lodging, so it was time for another shift of direction. Much to the delight of his father and grandfather, he applied for a course in clockmaking, and was accepted.

When he was first approached by the team behind *The Repair Shop*, his first thought was that the email was a scam.

The British Horological Institute course was run by a college in Hackney. Although Steve hadn't studied some of the relevant subjects at school, such as metalwork or drawing, he soon caught up and found that his natural ability to work with his hands stood him in good stead. But there was another equally valuable aspect to his education. For Steve, then still only 16, the move from a small market town in rural Oxfordshire to a hostel in inner London was a real eye-opener, exposing him to people from all walks of life.

After college, Steve returned to Witney, and set up his own workshop next to his father's. By then, his grandfather, also an horologist, had retired and spent a few hours every day overseeing his work, offering suggestions and generally extending his training. But jobs were hard to come by. Then, one day, Steve took the plunge and placed an advert in *Country Life* magazine. The only reply came from the projects manager of Trust House Forte, who commissioned him to service the clocks in all the group's hotels. From that moment on, he never looked back.

As for the solitary nature of the work, an aspect of clockmaking that had never appealed to Steve, the solution to that came from a surprising quarter. Following a visit from the station officer of the local fire station, Steve found himself invited along to drill night. What he learned was intriguing, and before long he had joined the retained fire service. (Retained firefighters carry on working at their normal day jobs, but are called up to incidents as and when they are needed. It's far from a back-up service, but an essential part of the front line of emergency workers.)

Working for the retained fire service gave Steve the human contact that he'd missed, together with a fair share of risk and adrenalin. After 30 years combining clockmaking with firefighting, and 12 of them as the station commander at Witney, Steve left to concentrate on his business, which quickly expanded. These days, there are 12 members of staff.

When he was first approached by the team behind *The Repair Shop*, his first thought was that the email was a scam. Luckily, his partner persuaded him otherwise and soon Steve had not only joined the show, but was instrumental in getting his sister, Suzie Fletcher, the leatherworker, to join him in the barn.

The environment in the barn makes it super difficult to repair small items like watches, which is why, with a few exceptions, we tend to concentrate on bigger mechanical things. A watch repair workshop should be laboratory clean – even a small bit of dust can be a problem.

Many of the clocks that come into The Repair Shop haven't been serviced for decades. Common problems include basic wear, dirt, grime, broken mainsprings, damaged cases. The mechanism may be jammed, or someone might have attempted a repair without really knowing what they were doing. Sometimes you have to make parts that are missing. The vast majority are easy to fix. If it's an ordinary clock with a normal mechanism, I could almost take it apart and put it back together again blindfolded. After 50 years, you build up a touch and a feel. That's not to say there aren't problem clocks – every horologist has encountered a fair few of these. One of the ones I worked on in the barn was made of bits of other clocks, which was far from straightforward.

The emotional side of the work really appeals to me. Clocks are special possessions for many people – there's that whole tradition of being given a clock on retirement, for example. And like jewellery, they tend to get handed down in the family, bringing with them their own associations.

People have to look after them, wind them up, and constantly check them to tell the time, so there is that human connection. When they come to pick up their repairs, many of my customers say that it's been dreadful to be without the tick of the clock in the house – they're almost like living things.

Some conservationists believe that ancient clocks belong in a museum and should be left untouched in case more damage is done. But I believe that anything that is supposed to work, should be repaired so that it still works.

I do love a challenge. I love learning things. Working in collaboration with the other experts in the barn always teaches you something new. Some of the most satisfying projects I've worked on in the barn are objects that I would never have come across during the course of my normal business. A case in point was the Wimshurst machine, also known as the "lightning machine", a fascinating piece of historical medical technology, which was a real privilege to work on.

Most mechanical people love their tools, and I am no exception. My favourite, favourite tool is my grandfather's old ratchet screwdriver. It's very simple, it's not worth anything, but it's very precious to me. I must have hundreds of old pliers, some of which I've adapted to do a specific job.

It's become something of a trademark of mine, but I couldn't be without a selection of glasses in different strengths. They're not prescription, just readers, but they enable me to cope with the fine detail. They're scattered all around my house and in my workshop. For really close work, I use a loupe.

I'm glad to say that my own kids have shown none of the same reluctance to get involved in the trade that I once had. My 21-year-old son Fred, named after his great-grandfather, is an apprentice clockmaker and has been down to the barn to help me out with a job, which makes me very proud.

Lightning Machine

Making the sparks fly

This rare scientific instrument, invented by the owner's great-great-grandfather, James Wimshurst, in around 1885, is properly known as a Wimshurst Machine. As well as being a treasured heirloom, it once made a key contribution to medical science and early X-ray technology, in recognition of which James was made a Fellow of the Royal Society in 1898. James made 90 such machines; this one dates from 1885 and has been in the family ever since.

The instrument features 12 glass discs, which rotate in different directions when the handle is repeatedly turned, generating static electricity via metal brushes that sweep against lead strips on the discs. Once sufficient static has built up from splitting the positive and negative charges, it is discharged through the spherical electrodes at the top in the form of a lightning bolt, hence the machine's nickname.

However, the years have taken their toll on the instrument; elements are worn and broken and it no longer works. Nick, James's great-great-grandson, is hoping that it can be repaired so that he can see the famous lightning bolt for the first time in his life and possibly loan the machine to a museum.

For horologist Steve, restoring such a notable Victorian invention is a dream of a repair job – and a daunting one. His first, delicate task is to disassemble the machine, carefully removing the glass discs. In the process, he discovers that a disc is missing and three others are either cracked or broken. All four will have to be replaced and 80 new lead strips made for them. The brushes need straightening and untangling so they touch the lead strips in the right places. Meanwhile, the wooden case makes its way over to Will Kirk's workbench for cleaning, waxing and polishing.

When the replacement glass discs arrive in the barn, Steve coats them with a layer of shellac, which serves as an insulator, and prevents any static electricity from transferring itself across the surface of the glass. Then it is a question of making the replacement lead strips. For these, Steve uses thick roofing lead, which he passes through a rolling mill to thin the metal – a process rather similar to making fresh pasta. These are embossed and

cut out to the right shape, then stuck in place on the new glass discs with shellac, using a broken glass disc as a template for positioning.

It's another heart-stopping moment when Steve comes to reassemble all the parts and fit the glass discs back into the newly refinished case. Next, he aligns the brushes with the strips, and slots the electrodes on their poles back into place. The original plaque on the machine read 'Wimhurst' and, as a final touch, this has now been replaced by one with the correct spelling of the family name. The instrument is ready to be returned to Nick – the question is, will it work?

Few handovers fizz with so much anticipation. Nick is not only keen to preserve an important piece of family heritage; he's also desperate to see that dramatic lightning bolt, with which his father once wowed his classmates at school. As he turns the handle and the bolt crackles between the electrodes, it's hard to tell who is more delighted: him or Steve.

Bringing back Nick's great-great-grandfather's invention is a lightning-bolt moment for both Nick and Steve.

A Clock Called 'Old Oily'

Bringing a giant timepiece back to life

Our expert horologist Steve Fletcher has seen a few things, but never before has he come across a 10-foot-high wooden clock designed to look like an oil rig.

This wondrously inventive timepiece has been brought into the workshop by Andrew and Monica Norton, from Botcheston in Leicester, who introduce the colossus by its affectionate nickname, Old Oily.

And so Old Oily's fascinating story unfolds. It was built by Monica's father, Ronald Godfrey Woodford, the son of a master cabinet maker. Ron was a cabinet manufacturer too, but he was also an inventor and from the age of 14, learnt the art of horology from an uncle.

Ron's first patented invention was a portable projection screen. This proved so successful that he ended up employing 15 men at his Harmony Cabinet Company in the market town of Earl Shilton, Leicestershire. Then in the 1970s, Ron began to combine his cabinet-

makeing talents with his horological skills and became well-known for his amazing 'Woodford clocks', so gigantic he could climb inside them.

Between 1975 and 1987, Ron produced five individual timepieces: the original Woodford clock the, the Greek Temple clock, which has been with the British Horological Institute (BHI) for 35 years; the Afrormosia clock (Afromosia is a teak-type wood used in cabinetry); Old Oily, now with the BHI to go on public display; and finally, his Queen Anne clock, also there potentially to be restored by BHI experts.

Designed for a clock show at the prestigious World Trade Center in Dallas, USA, Old Oily took more than 2,000 hours to make, after many months of design and testing. The timepiece itself is supported by wooden oil derricks (cranes) and the hands are also miniature derricks. The weight is contained within a wooden oil drum and the counterweight is a wooden replica oil boring tool complete with three individual carved wooden cutting gears. Ron had to patent the unique escapement mechanism - it works in reverse compared to an ordinary clock. Old Oily was designed to showcase Ron's unsurpassable skills. He stayed in Dallas to oversee it during the exhibition.

Ron wanted it to be seen, Monica says, in a place where it could be admired. Old Oily was eventually put

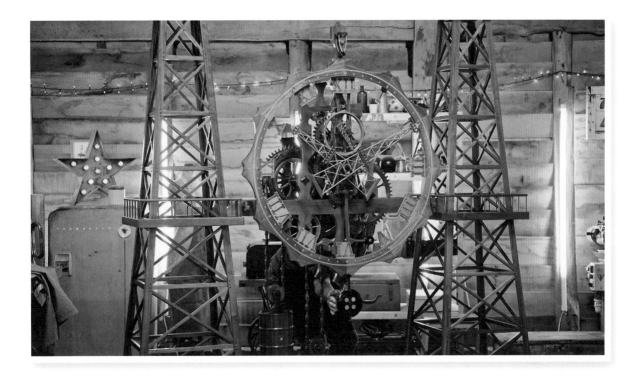

into storage with Ron's other clocks. It was becoming impossible to maintain them all. Monica remembers her dad nailing down the lids on the packing crates and saying, 'that's it now, finished'. The lid of the crate containing Old Oily hasn't been opened for more than 30 years.

Monica's warm, funny and caring dad passed away in 1990. Monica says seeing the workings of the giant clock laid out makes it seem like her dad has just gone out for a mug of tea and that he's going to come back into the room any minute. She fears Old Oily may have suffered from damp in storage. Andrew worked with Ron on his clocks for seven years but admits he lacks the restoration expertise required.

Old Oily needs the help of a master horologist. Steve Fletcher is just the man for the job, and he enlists the help of his son and apprentice, Fred.

Monica was right about damp; Steve can't believe how rusty the bearings have become. He orders replacements and turns to restoring the wheels and woodwork. Damp has penetrated underneath the lacquer and turned the finish white. Steve uses a linseed paint reviver to protect the wood.

When Steve and Fred have reassembled Old Oily, Fred prepares the weight – in the shape of an oil drum - so the clock can function again. Steve carefully adds it to check Old Oily will tick – it's the moment of truth, he says. When the wheels start turning and the clock does begin to tick, Steve is relieved; he's never, ever seen a mechanism like this before, he admits.

When Monica and Andrew return, Monica is shaking with emotion. It looks so beautiful, she says. Ron would be so proud and thankful, Andrew adds. Steve concludes it has been a labour of love; he has so much respect for Ron and the work he put into a timepiece that has stood the test of time.

Monica says that seeing Ron's clock back in working order and looking as it should is like having her father in the room with her.

Vintage Hearse Lamps

Shining their light once more

Returning from her travels around Europe in a 1972 camper van, Beth Haines stayed with her parents at their cottage in a small village in North Wales, enjoying the opportunity to reconnect through family history. As she and her parents chatted, talk turned to two, now-battered, vintage cast-iron carriage lamps thought to have been made in the 1830s. Up until World War II, these striking beacons had taken pride of place on the front of the horse-drawn hearses belonging to the family's funeral business, eventually owned by Beth's grandad, Charles, known as Ray. When her grandad passed away in 2011, her dad, Simon, placed the lamps in the loft for safekeeping.

Beth explains that her dad is passionate about family history and she loves to hear his tales of the family from before she and her brother were born. The lamps are special to Simon because, although he became a forester and moved to Wales rather than continue in the family business, they represent a link to his own past and his family's identity.

His dream, Beth says, would be to mount both lamps, as they were originally, in the porch at home, so they would be protected from the elements. They would then light the entrance to the house on gloomy days and spark conversations between family members and visitors. Beth's mission is to find a way of restoring these cast-iron heirlooms for her dad.

At some point, Beth's grandad had converted the lamps to electric, and hung them outside in the elements. Pitted with rust now and with flaking paint, their surface scratched and marked, and with a couple of dents, they've endured a lot of weather damage over the years. There are also no burning wicks now, and one glass panel has a significant crack down the middle.

Grainy black-and-white photographs show that the lamps were polished to perfection and still in use in 1936. However, Beth believes that they were retired during the war, along with the carriage horses, which were sent to rural East Dean from their stables in the middle of Eastbourne when air raids hit the coast. Following the war and the growing popularity of motor cars, the horse-drawn hearses were never returned to work, and the hearse lamps never used again.

Horologist Steve Fletcher, who has previously worked on an oil lamp in the barn, is confident he can restore them back to their original glory. He enlists the help of restoration expert Dom Chinea, who begins by stripping out the old electrics before sandblasting the metal. He then puts in the new lamp fittings and wicks before testing them to make sure that they work. Together, he and Steve devise new brackets so the lamps can be fixed securely to the wall.

Beth, who now lives on a remote Scottish island, where she is part of the land and livestock team, is fizzing with excitement when she returns to the workshop to discover how Steve and Dom have got on. She loves how the lamps now shine as they once did – Dom has relished applying a new coat of pristine gloss black paint and says that it's the best part of any restoration.

And Beth is in for a surprise. The family story always held that the hearse lamps were lit by oil, but Steve discovers that they were designed for candles. He then devises a special spring handle mechanism so that candles can be used once more. Beth is delighted that Steve and Dom have gone the extra mile to restore the lamps so expertly to their original working order.

Beth's wish was to have the lamps restored for her dad, and Steve and Dom have done just that – and more.

the
ceramics conservator

Kirsten Ramsay

Kirsten Ramsay, The Repair Shop's ceramics conservator, has over 25 years' experience in the trade. Specializing in the repair of oriental and European ceramics, she works with all types of ceramics: porcelain, terracotta and earthenware, as well as glass, painted plaster, alabaster, enamel, cloisonné and mosaic. Her clients range from museums and auction houses to antique dealers and private collectors.

Unlike some of the experts on the show, who were drawn to a particular craft from an early age, or who have had family members involved in the same business, when Kirsten left school all she knew was that she wanted to work at something arts-based. At college, she took a course in display and exhibition design, which involved technical drawing, graphics and shop window display.

After college, Kirsten moved to London and worked for an exhibition design company for a number of years, and then for Harrods, creating window displays. But something about the work left her feeling unfulfilled.

Working in the barn alongside the other experts has brought a whole new dimension into play. The exchange of knowledge and techniques, as well as the collaborative nature of many of the projects, provides endless scope to learn new things.

It was while taking an adult education course in ceramics restoration that the seed of her future profession was sown. When Kirsten overheard Penny Fisher, a ceramics conservator at the British Museum, explaining to other students that she was working on some ancient Egyptian pots, she couldn't believe such a job existed.

Everything finally fell into place after Kirsten moved to Brighton and heard about a postgraduate conservation course at the prestigious West Dean College. When she applied to do the course and was accepted, she found that ceramics conservation was a perfect fit. It appealed to her on so many different levels – as a hands-on person she relished the opportunity to use her practical skills, but the work also gave her the chance to express her artistic side and indulge her love of colour.

During her training, she completed internships at the Victoria & Albert Museum and at Brighton Museum. After graduation, she went to work for the British Museum in the same department as Penny Fisher, the woman who had sparked her original interest in conservation.

Since 1995, Kirsten has run her own ceramics conservation business in Brighton, working for clients from all over the world. Like many conservators and craftspeople, her work has tended to be a solitary endeavour. Working in the barn alongside the other experts has brought a whole new dimension into play. The exchange of knowledge and techniques, as well as the collaborative nature of many of the projects, provides endless scope to learn new things.

It's strange because my mother says I wasn't particularly patient as a child. I wasn't keen on doing jigsaws, for example, but in my profession I do become genuinely absorbed in what I do, which can involve piecing together lots of fragments, rather like doing a jigsaw in 3D. I get completely immersed.

One of the objects that came into The Repair Shop was a ceramic ornament depicting a bunny's cottage, treasured because it held memories of a happy childhood. It arrived in a box, smashed into something like 60 pieces, and some of these had become further damaged in transit. At first I wondered how on earth I was ever going to get it back together – my table was literally strewn with fragments. It was especially daunting considering the smaller base would have to be strong enough to bear the weight of the larger top part or superstructure. But I managed in the end. For many people we represent the last chance they will ever have to get their object repaired – it would be simply too time-consuming to be cost-effective for the average restorer.

One of the treasures I adored working on were unfired clay figures sculpted by a teenaged boy called Andrew when he was at boarding school. He was clearly a talented artist and had made the artwork for the equivalent of his A level. Tragically, he died in an accident, and afterward the school returned the sculptures to his family. For years they were proudly displayed but they were damaged

and then put in a drawer until his sister, by now an art therapist, thought to bring them to us. When the pieces arrived in the barn they were looking a bit tired and tatty, and parts had broken off. But you could still see that they were so beautiful and expressive.

You are close to the mind of another maker when you work on objects like that. The connection, the closeness of thought, really moved me. You can admire a beautifully made object and the skill required to create it, but it won't necessarily move you in the same way.

When I was employed as a conservator at the British Museum, I remember working on an ancient Egyptian mug and noticing the mark where someone had pressed their thumb to attach the handle. This was an object made in 3500 BCE and yet I could still see the thumbprint of the maker. That connection through time is magical. People have been making ceramics and baking them in the sun for millennia, and I find the social history absolutely fascinating.

With any repair that comes in, it's basically your own experience that tells you how long it's going to take. Proper cleaning, for example, for which you need to build in time for really studying and assessing the damage done to an object, can be a very slow process. You might be removing dust and grime as well as traces of adhesive and filling material from previous repairs – sometimes an old repair makes a task more challenging than a recent clean break. Even the adhesives we use can take four or five days to cure. If it was too cold in the barn, I'd sometimes put adhesives on a hot water bottle overnight.

Ceramics conservation brings together so many different elements that interest me. I remember doing pottery and clay modelling at school and absolutely loving it. And I've always loved colour, which comes into the picture when I'm mixing pigments to retouch or colour-fill, to give an invisible or discrete/sympathetic repair.

I'm constantly learning. Working at the barn has really brought that to the fore. I remember that first week when we first started working together, and were complete strangers, the production crew noticed how interested we all were in what the others were doing. That hasn't stopped. We are always fascinated by each other's different skills and techniques, and wondering whether they could be usefully transferred to our own work.

Garden Urn

Picking up the pieces

Adam Steward grew up by the sea in Devon, but his father's family hailed from the proud steelworking town of Stocksbridge, near Sheffield, where his great-grandfather, John, was a successful house builder. John also owned a brickworks and a pipeworks, with a clay mine beneath.

Although his own home was hundreds of miles away in Devon, Adam's regular visits to industrial South Yorkshire with his sister Rachel helped to shape his life. He loved to see his grandparents, and eagerly anticipated the hugs and kisses and the sight of their gardens, where two family heirloom urns took pride of place.

And this is why Adam is here today. He can't think of his grandparents' house without seeing the twin salt-glazed earthenware garden urns that stood almost a metre (3ft) high in their immaculate front garden. The urns, made in 1930s Art Nouveau style for John Steward by his own men, using clay from his own mine, had been passed down to his son and family, and then to his grandson, Adam's dad.

When Adam's grandparents died, their son, also named John, inherited the urns. Shockingly, after residing in their new home for only two weeks, a retaining wall collapsed on them. One of the urns bore the brunt of the damage, which shattered the bowl into chunks and cracked and chipped the clay plinth. Adam's dad picked up the pieces and put them in his own workshop, and that's where they sat for 25 years, waiting to be repaired. Adam says his dad was always so busy helping other people and time ran out on him. Sadly, in 2018, he died from cancer.

Clearing out his dad's workshop has been so difficult, Adam says, so many memories and so many things left undone. But that's where he found the broken heirloom urn that has brought him here today.

Adam now has two grandchildren of his own. He would love to have both urns standing either side of his front door, to pass on the family legacy. He vows to make sure everyone knows how precious they are and where they have come from.

This is definitely a job for ceramics expert Kirsten Ramsay. She examines the shattered urn and sees that the Yorkshire clay of the bowl has fallen into large unusable chunks. Also, the intact urn still has a small connecting metal rod for support, but this is missing in its twin.

Worryingly, there's a plastic or glass fibre coating that's left a residue around the bowl of the urn – Kirsten thinks this might have happened if it had once been used for planting, and the paint is flaking off in the interior of the bowl.

When Kirsten gets the urn on her workbench she notes the orange peel-like texture – it happens in the kiln, she explains, creating a strong frost-proof ceramic. She makes a start removing the plastic or glass fibre coating. She is quite taken aback with the level of damage, suspecting that repairs may have been required before the wall incident.

First, she cleans the shattered pieces with her steam cleaner, discovering yet more damage. Clearly, the damaged urn has had a lot of historic restoration, with car body filler, among other remedies. She resorts to use paint stripper to break down previous repairs, before carefully reassembling the bowl, bonding and filling the chips with coloured filler.

Now the important part: leaving the bowl to cure so the repairs set.

Jay is concerned for the stability of any repair, especially with grandchildren around. He calls in silversmith Brenton West to assist with making a new sturdy rod fitting to secure the urn to the plinth.

The final test – carefully reuniting the pieces to complete the urn – is passed and, finally, it is ready for Adam to take home. Adam returns to The Repair Shop accompanied by his sister Rachel. When he sees the urn back in one piece, he is over the moon. He is so proud that his great-grandfather's legacy has been restored, and he can't wait to place the urns by his own front door. From now on, he says, he will have a smile on his face every time he comes home.

Now the urn is back in one piece, Adam is looking forward to telling the stories surrounding its history to his own children and grandchildren.

Buddhist Figurine

Fixing a fascination

Retired GP Mickey Adagra moved to England from Mumbai, India, in 1977, but he's kept a treasured ornament from his childhood safe for decades. He explains that it's an early 20th-century porcelain figurine of Guan Yin, the Buddhist goddess of compassion and mercy, from China, which belonged to his dear aunt and uncle, Maki and Adi. His aunt's grandfather used to trade with China for his work and on one of his trips he brought the figurine back with him as a present for her.

Maki and Adi had no children of their own and welcomed Mickey to their home, which was in a small town about 120 miles from Mumbai, every school holiday. They lavished all their love and attention on their nephew, who was like their surrogate son.

Mickey, an only child, looked forward to these visits as much as they did. They were the only holidays his parents could afford; his father worked on the railways but, crippled by painful bouts of rheumatoid arthritis, he was often too ill to work and was bedridden, which left his mother, a schoolteacher, the breadwinner.

The figurine, delicately painted in soft traditional Chinese tones of red and blue, and standing on a fish, has a secret. She's known as a 'dripping lady'. If you put water in the fish, turn the figurine upside down then stand it upright, the water drips from a flask in her hand straight into the fish's mouth. In Mickey's case, at least, that's what she used to do when he was a child. He was fascinated by this seemingly magical action and would beg his aunt and uncle to demonstrate the trick over and over again.

Sadly, the right-hand fingers and the tip of the tail of the fish are now broken off, as is the figurine's left hand with the water flask, and the water no longer flows. Accidents occurred when cleaning it, his aunt told Mickey, and repairs have been made over the years. Thankfully, his aunt kept the broken hand and passed it on to Mickey.

Mickey would pay Maki and Adi a visit every time he returned to India, which was nearly every year. When his aunt was quite elderly, she offered Guan Yin to Mickey, on condition that she remained in the family and wasn't sold or given away. Now Mickey wants to honour his aunt and uncle's memory and see Guan Yin in full working order again, but he's concerned that after all these years she simply won't be fixable.

His plan is to keep this special lady safe in his family and eventually bequeath the heirloom to his son, who is equally fascinated by the goddess but has never seen the water pouring from her hand.

This is a job for The Repair Shop's ceramics expert, Kirsten Ramsay. Her concern is that the crucial area of damage on the hand is exactly where the water comes out of a little flask. The end of this has been lost, so she will need to make it up.

She starts by using a steam cleaner, which is perfect for glazed porcelain, on the figurine's body, as well as the broken hand piece. It removes all the existing adhesive from past repairs, as well cleaning the hand holding the flask and inside the hole where the water used to drip through. She carefully applies adhesive to the outside of the broken hand, avoiding the hole, and attaches it to the figurine, and leaves it undisturbed to cure.

To make up the end of the flask and the missing fingers and thumb, Kirsten uses two-part epoxy resin, adding a bluish colour to match the background porcelain. When making the end of the flask, she inserts a pin to keep the water hole open then surrounds the pin with filler, shaping it as best she can. Once she's shaped the missing fingers and thumbs, she leaves everything to cure.

Mickey is so grateful to Kirsten for restoring his porcelain goddess and enabling him to honour his dear aunt and uncle's memory.

When Kirsten eventually removes the pin, she is relieved to find that the water channel has remained open. She is now ready to apply a glaze coating over the repaired fingers and thumbs and flask to mimic the glaze of the figurine. Again, she uses a two-part epoxy resin but this time it's clear. Finally, she makes up a new fin and colour-matches it to the original red-brown.

Mickey is very excited at the reveal – he hasn't seen Guan Yin working for over 50 years – and he can't wait to see what Kirsten has been able to achieve. He inspects the figurine closely, noting all the repairs that Kirsten has made, and reveals that never in his wildest dreams could he have imagined such a transformation. He can barely contain his excitement when Kirsten fills the fish's mouth with water and turns Guan Yin over and back. He is overjoyed to watch the water dripping from the flask back into the fish's mouth, and is transported straight back to his aunt and uncle's home when he was a small boy.

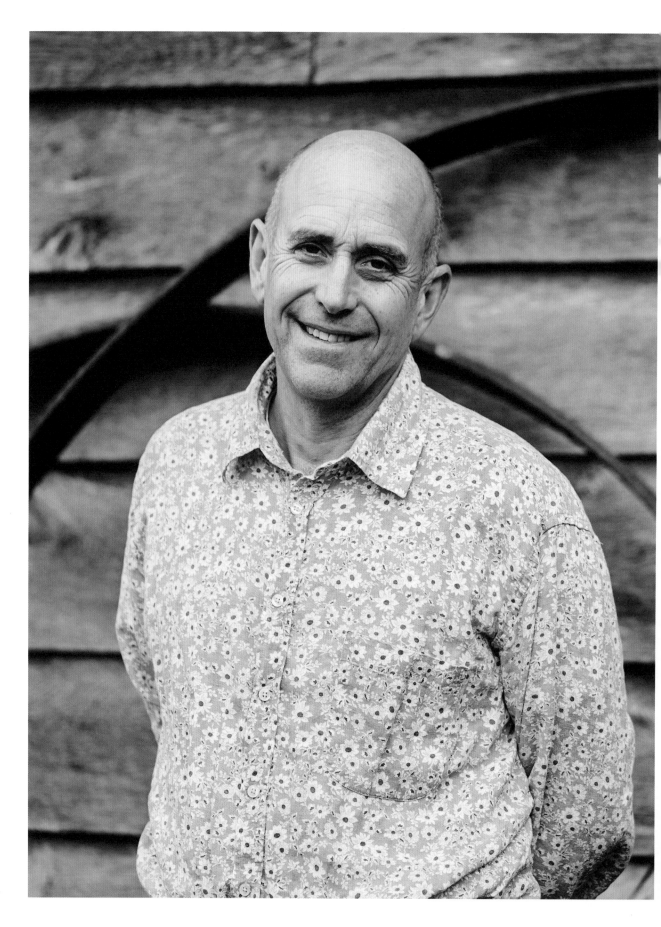

the
silversmith

Brenton West

The Repair Shop's resident silversmith and all-round metalworker, Brenton West, first became involved with the show in his capacity as an expert in 19th-century photography, when he was rung up by the producers and asked if he could repair a bellows camera dating from about 1890. During the course of the restoration, which involved cutting out a number of brass elements, as well as piecing back together a hundred different bits that had come unglued, he let it be known that he had 40 years' experience as a silversmith. Now a regular in the barn, he turns his hand to the repair of an incredibly wide range of metal objects, from Olympic medals to a seaside telescope.

Brenton's fascination with metalwork began during his school days at Bryanston, when an inspiring young teacher taught him to make a variety of items out of copper. Unusually for a public school, Bryanston has always had very strong craft departments – woodwork, pottery, sculpture and art, as well as metalwork – and Brenton, with his strong practical bent, was drawn to them all.

From the moment he arrived at the barn and soaked up the atmosphere of creative teamwork, he was determined to stay.

Deciding to pursue a career in silversmithing was a natural next step, and after school he applied to, and was accepted by, Medway College of Design in Kent, one of only four students taken onto the course that year. Here he was taught by some of the very best silversmiths in the country. Graduating with a good diploma, and with a unique set of skills under his belt, he set up a silversmithing business in Oxfordshire with a friend from college because jobs were thin on the ground. But so, too, were commissions and soon Brenton was looking for other work.

A self-confessed petrolhead, and with the skills that meant he could bend any kind of metal, he went on to work for a company that restored Formula One cars and classic cars, such as Ferraris and Lamborghinis. After a few years, he set up his own classic car restoration business and ran that for a while, before heading off in a completely new direction, which was property.

Brenton's route to The Repair Shop was a direct result of him taking a degree as a mature student at Plymouth College of Art and Design, where he specialized in pre-1900 photography, specifically daguerreotypes. Unable to afford a daguerreotype camera himself, he made one out of wood. Eventually he launched a website hiring out similar cameras he had made himself to television and film production companies – one of his cameras featured in the film *Effie Gray*.

From the moment he arrived at the barn and soaked up the atmosphere of creative teamwork, he was determined to stay. Realizing that the programme was unlikely to get all that many historic cameras to repair, he mentioned his background in metalworking and silversmithing. By chance, Will needed a piece of metal cut out for a desk he was repairing, and Brenton soon became an indispensable part of the team.

Everyone in The Repair Shop is basically working on things we might never have considered working on before. I'm quite handy – I can do joinery as well as metalwork, for example – but that is also true of the others. There isn't a steady stream of silver items for me to repair, so I will tackle anything from an old steamer trunk to beekeeping equipment. It's very rare that something can't be repaired. We all try to help each other out. If the minds in the barn can't do it, it probably can't be done.

Many of the metal items that come into the barn are tarnished, scratched, dented or broken. The first step is to mentally break down what needs to be done and decide on a plan of action. I generally polish first, so I can see exactly what the true condition of the object is – whether it's got any big dents, scratches or cracks. I use silver wadding polish on silver, and brass polish on base metals. Never use brass polish on silver plate, because it is much too aggressive; use silver polish, but you have to be gentle because if you are overzealous, you can polish right through the silver plate.

Any dents I will hammer out. People tend to think they have to be very careful with their silver items, but silver is just another type of metal; it just costs more than copper or brass. You can be quite aggressive with it.

If an item is deeply scratched, I will file it first, then go over it with increasingly fine gauges of wet and dry papers. Then I will polish it with Tripoli compound, a fine-powdered porous substance formed into a bar, and finish up with a jeweller's rouge.

Broken items often require soldering. And here's where you can run into a problem if there's been a home repair. People are often tempted to use lead or soft solder on silver because it's easy to work with. However, if you heat up lead solder anywhere near the temperature you need to silver solder, it will just burn a hole through the silver like acid – even a pinprick will. Some silversmiths won't even let you take soft solder into their workshops at all. If a silver item has been mended with soft solder, the only thing I can do is repair again with a soft solder (I use a tin/silver mix). In the barn, I solder on bricks that are pottery kiln shelves – one for soft solder and one for silver solder, keeping them strictly apart from one another.

A related problem can come when you are working on items made of white metal. White metal, which is the generic term for metal alloys such as spelter, pewter and Britannia metal, has often been used to make budget artefacts. It melts at a low temperature, which makes it easy to cast. But it can be a pig to repair. If you go near it with a flame, you can end up with a pile of liquid, shiny metal. Sometimes I will use glue rather than solder because it's just too risky.

To repair a silver-plated object, I may need to silver-plate it again using a brush plating kit. To electroplate using a brush plating kit, a solution of silver salt is applied to the brush and then you make a circuit between the brush and the item, brush the item evenly, and the silver molecules in the salt detach themselves and attach themselves to the surface.

Among my favourite tools are my hammers, some of which I've had for 45 years. They're like an extension of the hand. People like to pick them up because they're so tactile. Metalworking hammers aren't like ordinary hammers. They come in a range of funny shapes, designed for different jobs or to access different areas of a metal object.

There's such a good atmosphere in the barn, which is why I really wanted to stay. We all learn so much from each other. It's a cliché but we're really like a family. I'm quite shy and working here has really brought me out of myself and given me confidence. If you'd told me ten years ago that one day I'd be on the telly, I'd have said no chance!

Beekeeping Equipment

Keeping a legacy alive

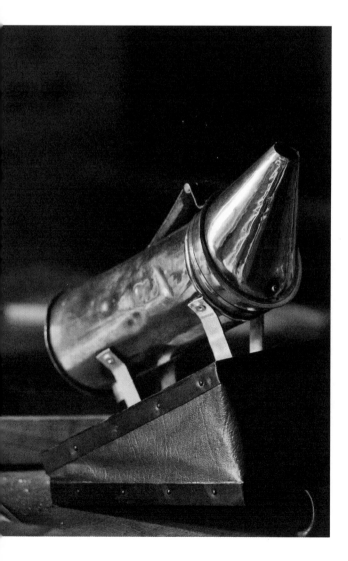

Silversmith Brenton West is buzzing when Vanessa Tyler turns up at the barn. As an amateur beekeeper himself, he's fascinated to meet Vanessa and learn all about the vintage beekeeping equipment she has brought with her from Cornwall: a honey spinner (also known as an extractor), which works by spinning the frames of honeycomb very quickly, separating the honey from its frame by centrifugal force and gathering it into a reservoir below, and a bee smoker with bellows. This is used to calm and clear bees from their hive so frames can be collected.

Brenton is even more pleased when she tells him the equipment belonged to her father, Alan, who had kept bees since Vanessa was a child, with five hives in his garden. When Alan was in hospital battling cancer in 2016, Vanessa made a solemn promise that she would take up the mantle of family beekeeper and ensure that his equipment was kept in use. As he lay in his hospital bed, he worried about his bees. Vanessa promised to take care of them and says she thinks about her dad every time she puts her bee suit on; she's taken the tag from his own suit and sewn it into her own, so he's always close to her when she is beekeeping.

The spinner's mechanism is technically working, but it's all rusted up, Vanessa says. The smoker isn't operating properly because the bellows leak and it's been pinned and mended so many times it needs a thorough overhaul.

Brenton is so excited to work on these fascinating pieces of equipment. He would love to have a vintage honey spinner of his own, he admits.

The smoker is first to receive his expert attention before he moves onto the more complex spinner. He strips it down, cleans the wood, checks the bellows and burnishes up the copper, in preparation for reassembling.

Next, he addresses the spinner and sends off the barrel, cage and meshes to be freshly tin-plated so they will be food-safe to make perfect honey. Disassembling the mechanism to free up all the parts, he's concerned that there may be holes in the bottom of the barrel, but it's fine. All he needs to do is remove the tap off the front. A bolt snaps – but it's not a huge drama, he can add a new one. The gears and other moving parts are removed to be cleaned and painted, in a deep green, the closest shade Brenton could find to the original paintwork.

As the paint is drying, Brenton turns back to the smoker. The bellows material is coming away from the wood, causing a very slight leak. However, before he tackles this, he's noticed something rattling around inside and discovers a loose bolt that has fallen off and been replaced by a screw. This is attended to, by refixing the bolt and holding it into place with a washer. He polishes the smoker's burner and reattaches its board to the bellows with staples, encircling the whole thing with the decorative band, which he'll repaint to smarten up.

Raiding Will's bench for some wood-shavings and adding a few balls of scrunched-up paper, he tests the smoker outside. It's working fine and Brenton is delighted; he says he would be happy to use it on his own bees now.

Back inside, he returns to the honey spinner. All the components have come back from the tin-plater, so he can reassemble it. It's a bit like a jigsaw puzzle, he says, but he knows how to put all the bits back in the right order.

Vanessa returns in the hope that Brenton's hard work will allow her to keep her final promise to her dad. She says she's nervous as she awaits the reveal; if the equipment is usable again, every time she tends to her

bees, she will hear her dad's voice in her ear and remember all the things he told her about the bees.

Her voice breaks with emotion when she sees the result; she hopes her dad can see his gleaming smoker and spinner. As she and Jay test the spinner and extract some fresh honey, everyone agrees that this very special connection has been restored.

Omani Chest

Rekindling childhood memories

The treasures that come into The Repair Shop are always freighted with memories. But few carry the weight of so much family history as this Omani chest, which is studded with brass nails. Mahbuba Abdullah bought the chest at a souk in Oman in the early 1970s.

Mahbuba was born in Zanzibar, an island off the east coast of Africa. She was brought up by her grandparents, and spent a happy, carefree childhood growing up by the sea. But all this changed overnight when a violent revolution overthrew the government in 1964. Mahbuba and her grandparents were forced to leave the country and all of their possessions behind.

For Mahbuba, aged 12 at the time, it was all a great adventure, but she remembers the worry etched on the faces of her grandparents as they confronted an uncertain future. Eventually, they arrived in the UK and settled in Portsmouth on the south coast.

Before the revolution, Zanzibar, which had long

been a cultural melting pot through its trading links with the Arab states, had been ruled by the Sultan of Oman. When Mahbuba, her husband and two children travelled to Oman to work in the late 1970s, memories of home came flooding back. In the winding alleys of the souk, she spotted this chest and remembered all the others like it that had been scattered about her childhood home. Traditionally used as dowry chests – places to store silks, jewellery and perfume – they also made irresistible hiding places for children. She bought the chest on the spot and had it shipped back to the UK.

By the time the chest arrived at The Repair Shop, it was in a sorry state. Over the years, house moves and time spent in storage had taken their toll. The brass nails that make up the decorative pattern were badly tarnished and a number of them were missing altogether; the original wood was bleached and dull-looking. The question was: could the chest be restored to anything approaching its former glory so that Mahbuba could pass it on to her daughter, Nadia, as a family heirloom? Nothing else – not even a single photograph – remained of her Zanzibar childhood and the chest was the closest thing she possessed to her own missing family heritage.

With both the brass nails and wooden finish needing attention, the project was a natural collaboration for silversmith Brenton West and woodworker Will Kirk.

First up was Brenton, whose initial challenge was to clean years of tarnish and verdigris from the nail heads using successively finer grades of wire wool. It was a meticulous and painstaking task that took two days.

The next step was no less daunting. According to Brenton's reckoning, over 80 brass nails were missing, and these would have to be made by hand. To do this, he used a punch and a block, placing a flat brass disc into the appropriately sized recess in the block and hammering it into a dome shape. He then soldered the domed head to a nail, and repeated the process at least 80 times.

By the time the chest was passed on to Will, the brass work was gleaming and immaculate, all the nails were present and correct, and the polishing had cleaned years of grime off the wood, too. What remained was to bring out its natural colour and grain. Using a soft pad and oil, Will built up coats and coats of finish until the wood could not absorb any more. Brenton supplied the finishing touch: two brass handles to be fixed to the sides of the chest.

When they came to collect the chest, both Mahbuba and her daughter, Nadia, were completely overwhelmed by its transformation. For Nadia, who remembers the chest being around ever since she was a little girl, it is something to cherish and pass on in time to her own daughter, Sabria. It is also a reminder of her relationship with her mother. For Mahbuba, it represents a vital link to her childhood in Zanzibar.

Brenton and Will restore the damaged and tarnished chest for Mahbuba, a poignant reminder of her childhood home, which she had to flee when she was just 12 years old.

the art conservator

Lucia Scalisi

Lucia Scalisi describes her decision to become an art conservator as a 'real lightbulb moment'. Training as an artist and with a degree in English and Education, she had always been very drawn to painting and the visual arts, but knew how difficult it would be to make a living as an artist. She was volunteering at a local museum and art gallery, helping to put on exhibitions, when a casual conversation with a friend of a friend mentioned someone who was training to be a conservator. The next day Lucia called Sheffield City Art Galleries to enquire further. Once she visited their conservation studio in the city centre, the spark was lit.

But Lucia also had a love of travelling and it was a couple of years, mostly spent in the US, before she won a place on the art conservation course at Newcastle University, one of only three such courses in the country. Along with her passion for paintings, her knowledge of chemistry also stood her in good stead, because the training was – and still is – science-based.

Lucia has been The Repair Shop's resident art conservator since the very first series, carrying out some astonishing restorations, where years of grime are cleaned away to reveal the stunning artwork beneath.

Before Lucia had finished the art conservation course, she had secured a job at the Victoria & Albert Museum. One week after she left Newcastle for good, she was down in London. At that time, the Victoria & Albert Museum had a large conservation department covering a huge range of disciplines, from books, sculpture and stained glass to plastics and furniture. Lucia was employed in the painting section for the next 12 years, which effectively extended her training.

During that time, she was seconded to work on a five-year project for the Calcutta Tercentenary Trust. This trust was set up to rescue deteriorating European paintings and artworks housed in the resplendent Victoria Memorial Hall, built in the 1920s when Kolkata (then Calcutta) was still India's capital. Once Lucia had left the Victoria & Albert Museum to set up her own studio in London, she carried on working on that project, travelling out to Kolkata for two to three months a year. Eventually, with funding from the Getty, she wrote a training course for future conservators on the collection, which includes one of the three largest paintings in the world –

incidentally, one that Lucia herself has worked on.

Lucia's subsequent career has neatly dovetailed her love of travelling with her role as a conservator. As well as studio-based commissions, she has been employed on projects from Tbilisi to Beirut, among many other far-flung destinations, and relished the opportunity to meet many interesting people in museums and art institutions around the world. Closer to home, she has been The Repair Shop's resident art conservator since the very first series, carrying out some astonishing restorations, where years of grime are cleaned away to reveal the stunning artwork beneath.

The barn can be a very challenging environment, especially in the winter if you are waiting for something to dry. I have a particular problem with varnishes – sometimes, they just take ages to go off. Paintings don't like the cold and damp, either.

The repair jobs that come into the barn for me have a range of problems. Usually the paintings are dirty and sometimes they have suffered physical damage in the form of rips and tears. The paint surface may be flaking. Often, they have been dabbled or tampered with – someone might have overpainted a section or added an extra layer of varnish. Taking that layer off might reveal even more damage, or great original paint – you can never tell.

The first step is to carry out a cleaning test, going over a small portion of the surface very carefully with a wetted swab in order to identify the solubility of the dirt. This can vary depending on whether there's an underlying layer of varnish. If there isn't a layer of varnish underneath, that's a whole new problem, because it means the paintwork has had no protection and the dirt will have become embedded into the surface. Another complication can arise if a varnish has been placed on top of another dirt layer. The cleaning test also allows me to have a close look at the painting and determine what needs to be done.

Cleaning a painting makes for very dramatic "before" and "after" shots. But generally it can't be rushed. Currently in my studio, I have a "six-footer" that I'm cleaning. It's been there a year and I'm only three-quarters through. In the barn, however, we do have to keep the time frame more manageable, which doesn't mean skimping on the repair, but being careful about which paintings we choose to restore.

There are a range of solvents I can use, depending on what I am trying to remove; sometimes I make my own. Most of the paintings that come into the barn have natural resin varnishes, which tend to be more soluble. Then it's a question of going over the surface of the painting very carefully with swabs. Different pigments have different solvent sensitivities. Those that are very glossy, where the pigment is mixed with a great deal of resin, need especial care. One of my specialist areas is pigments and the working practices of artists, the materials they used during different periods, which helps me work out how to proceed with the cleaning and conservation.

Art conservation is a small market, and most tools and materials tend to be borrowed from other industries. My basic tool kit consists of brushes, swabs, cotton wool, acid-free tissue, solvents, adhesives (water-based, acrylic and heat-seal), varnishes and pigments. I've learned from my time working in developing countries that you can do a lot with very little and still get great results.

Varnishes are complex things and I usually make my own. Their qualities vary, not simply in terms of appearance – for example, matt or glossy – but what they offer to the surface of the painting from a consolidation point of view.

I also have a thermostatically controlled spatula, which I use for flake-laying. First, I'll lay acid-free tissue over the flaking area, feed adhesive through it and dab with my finger to soften the flakes and get them to lie down. Then I'll go over it again with the spatula, which dries out the adhesive and keeps everything in place.

For retouching, I use Winsor & Newton series 7 brushes, sizes 00, 0 and 1. Handmade from sable hair, they are fantastically springy, with a good flow of paint right through to the tip. Nothing compares to them – and I've tried everything.

On the whole, I'll only retouch an area where there's been a loss or tear and I've had to do a filling. I never retouch on top of the original paint. Before filling and retouching, I will add a layer of varnish, to isolate anything I put on the original surface. After filling, I'll apply an opaque base coat, usually acrylic, watercolour or egg tempera, then mix my retouching colours – dry pigments bound in a transparent synthetic resin.

No one retouches with oil paint. You may get a really good colour match at the beginning, but oil is a very complex medium and takes about a year to dry. During that time it goes through a cross-linking oxidation process, which means that it will change colour. Modern materials tend to be much more stable and are certainly less toxic. Pigments mixed with synthetic resin don't cross-link, which means the layer can always be removed at some point in the future.

Art conservation is a vocation, not a job. It's meditative and it's slow. It takes a while to get your eye in and really focus. Some days, when you're in that zone and the work is really flowing, it's the true meaning of mindfulness. On other days, it's just not going to happen and you have to walk away for a little while.

Queen Henrietta's Portrait

Revealing a hidden story

When she first spots this beautiful – but battered – portrait, painting restorer Lucia Scalisi is taken aback. It is an extremely rare portrait of Queen Henrietta Maria, in mourning for the loss of her husband, Charles I, who was executed in 1649.

The 'Henrietta' painting has been brought in by Howard Bird from Hertfordshire. He was given the portrait by his neighbour Eric, when he passed away in 2016. Eric had acquired the painting from an art dealer friend in the 1960s. Howard, Eric and his wife Ruby

were all great friends, so close that in 1998 they swapped homes: Howard's apartment was more wheelchair-accessible, which suited Eric, who had contracted polio at the age of six and lost the use of his legs.

It would mean so much to Howard if the portrait could hang again in the house Eric and Ruby once lived in themselves. He wants to fulfil Eric's wishes, to get Henrietta back to the beautiful woman she is. Unfortunately, Eric had become seriously ill before he had the opportunity to have the painting restored. When it arrives, it appears almost black from dirt and smoke, as it was hung close to a coal fire.

Lucia says that there could be a lot of paint underneath Henrietta's portrait, and also – potentially – damage. She won't know until she begins the painstaking process of restoration.

First, she removes the painting from its frame. The first cleaning test, with distilled water and a drop of ammonia, will just remove surface dirt.

Next, she does a varnish cleaning test with hydrocarbon solvent; yellow lumps of varnish curl up, to reveal Henrietta's rosy-pink flesh beneath.

Happy, Lucia begins the full surface and varnish clean. Having carefully removed the dark surface dirt, overpaint and varnish, she can finally assess the painting's original condition. The background had been over-painted in a black paint; there is no space in its place. Even Henrietta's dress had been painted black, but Lucia's work reveals it as a dark grey satin, with red flashes in the sleeve.

The high-saturation varnish she will use – chosen because it emphasises colours and will work well with the retouching process - is the first modern varnish the painting has received. This coat of varnish goes on before anything else, to provide an "isolating layer" between the original paint and the retouching work.

Being so close to the painting, Lucia starts to see things of interest. One theory she ponders is that it may be French in origin. However, there is now a three-dimensional aspect to the figure which enables Lucia to make a startling discovery; Henrietta is pregnant. Her hands are placed demurely across her belly. At this period of portraiture, the pose was usually an indication the subject is with child.

Removing extensive overpaint from her dress reveals that it had been altered to cover Henrietta's maternity and change her into a widow. For political and dynastic reasons, she had to have her child in England, meaning that the painting could not be French.

Dating the painting is difficult, but it is known that Henrietta's last pregnancy, at the age of 35, was in 1644. Her husband never saw the child because he was imprisoned before she gave birth. This may explain the mourning accoutrements.

Henrietta has a wedding ring on a black band at her left wrist, and a miniature of a man, believed to be Charles, at her breast. Lucia says that these are later additions – the paint and handling of them are not only different, but they are out of proportion to the rest of the portrait.

She fills and carefully retouches the surface to bring Henrietta back to life. Finally, she revarnishes the painting to give it a beautiful finish before fitting it back into the

frame. Lucia suggests that the portrait is possibly from the Sir Peter Lely workshop. Lely was court painter to Charles I, and he would have painted Henrietta's face, with everything else done by the workshop, standard practice at the time.

Howard is joined by his husband Denes for the big reveal. They are so pleased to see 'Henrietta' beautiful again. It has been such a fascinating journey, starting with that first step toward The Repair Shop for restoration and finally learning that Queen Henrietta was painted when she was pregnant. Howard says Eric would surely be looking down in admiration.

It is only because Howard brought his painting in to The Repair Shop for restoration that he learns that Queen Henrietta was painted when she was pregnant. His dear friend and neighbour Eric would have been delighted at the discovery.

Portrait of a Little Girl

A moment in a family's history

When Sophy Bellis was six years old in the late 1960s, her mother, Barbara, asked renowned artist Roger Hampson (1925–1996), associated with the realist Northern school pioneered by L.S. Lowry and Harry Rutherford, to paint her portrait.

He was known more for his atmospheric depictions of pit heads and gnarly street corner characters than likenesses of children, but luckily, Mr Hampson, who was principal at Bolton College of Art, lived just up the road from Sophy and her family. He captured her in a grey smock dress sitting on a 1960s orange chair.

Sophy's portrait takes her back to the warmth of her childhood. She says that Mr Hampson captured her really quite brilliantly, even though he said he hadn't quite got her mouth right; she was so excited during the sittings she wouldn't stop talking.

Her mother adored the result. It hung on the wall and was much admired, always providing a talking point when Sophy, who lives in Shrewsbury, Shropshire, came to visit. But over the years the surface began to flake. Barbara was desperate to find some way of stopping this, and intermittently would mention trying to find someone to look after it for her.

Sadly, Barbara developed dementia. Although she found it difficult to create new memories, Sophy says that her long-term memories associated with the painting would always provoke the strongest reaction. This made the painting even more special for Sophy, and her mum; it was a vehicle for them to connect until the very end.

Now, Sophy is desperate for painting conservator Lucia Scalisi to see what she can do to slow down the passage of time, to cement an important link to her past and bring the painting back to life, so that she can put it on the wall and think of her mum.

Lucia says that she absolutely loves the 'intimate' portrait of Sophy with its wonderful 'opalescent colours'. However, there is a significant amount of damage, she warns. The oil paint is flaking and curling up off the canvas. Lucia soon discovers the root of the problem. The painting has been done on a reused canvas. This is not an unusual practice, but Lucia feels that the artist had scraped off a previous painting done on the back. The process has left the canvas stiff and unabsorbent, a major factor causing the paint to flake, Lucia believes.

Her approach to this challenging restoration job is first to consolidate the existing paint with an acrylic-based consolidator. This is overlaid with acid-free tissue before Lucia uses her heated spatula to encourage the adhesive in the consolidator to flow and the paint to soften. This process will help to save as much of the lifting paint as possible.

When she has finished her painstaking task, Lucia

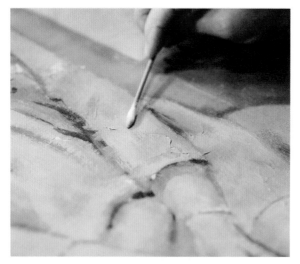

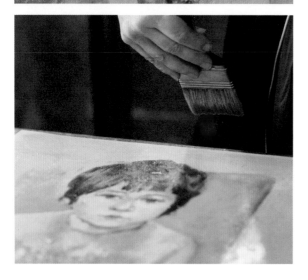

removes the tissue paper piece by piece. She's delighted to discover that there are far fewer paint losses than she originally thought; with a re-used canvas, you never know quite what to expect, she says.

She next applies a thin coat of varnish to act as an insulating layer. When the varnish has dried, she then starts delicately filling the paint lossess with an acrylic gesso filler – using a dental tool – and gentle retouching with pigment.

As she works, Lucia reveals a secret of her own. She has never before worked on a portrait where the subject is still alive and able to comment on the result.

And when Sophy does return to collect her portrait with her son, Harry, who will eventually inherit the painting, she gasps in astonishment. She says she cannot remember the precious portrait ever looking as good – and is looking forward to taking home this poignant depiction of her childhood self.

Sophy knows that every time she looks at the restored portrait of her as a child, memories of her mum will come flooding back.

the upholsterer

Sonnaz Nooranvary

Skilled in both modern and traditional types of upholstery, Sonnaz Nooranvary can turn her hand to the repair of anything, from an antique button-backed leather armchair to a mid-century modern design classic. She is one of the few experts in the barn who still runs her own business; she first learned her trade at the luxury motor yacht builders Sunseeker. For Sonnaz, however, finding her true metier was not a straightforward path.

Born in Iran, Sonnaz grew up in Hampshire, then the family moved back to Iran for a couple of years before returning to Britain. She was always creative as a child, happiest when she was making or drawing, taking inspiration from her mother, who made all the family's duvet covers, bedclothes and curtains. Left to her own devices, Sonnaz might have pursued her creative bent sooner were it not for her father, who, like many Middle Eastern parents, had decided that the law, medicine or dentistry were the only suitable professions for a woman – either that or marriage and children.

But less than a month before Sonnaz was due to take up her place at college in Southampton, where she would be studying law and sociology, among other subjects, she knew she was about to make a mistake. With her mother's encouragement, and the cooperation of the college, she changed to a selection of arts-based courses. This proved to have its own drawbacks, as a large part of the final grade was written coursework and Sonnaz struggled with dyslexia, which had never properly been diagnosed.

The desire to do more furniture-based work, to make bespoke items for the home and to have a direct relationship with clients led her to quit her job and take the momentous step of setting up her own business at the age of only 25.

The solution presented itself when she came across an advert for an apprenticeship at Sunseeker. This prestigious international brand, based in Poole, Dorset, was offering 25 places, one of them in upholstery. Sonnaz, who had a hankering to do sculpture, applied for the upholstery apprenticeship and won the place despite some very stiff competition.

She spent eight years at Sunseeker, the first four of them as a paid apprentice. She accepted every opportunity she was offered: as the company's first female apprentice, first female in the upholstery department, and first female in the design office, and she also worked in soft furnishings and development. All the while she was hungry to learn more skills.

The desire to do more furniture-based work, to make bespoke items for the home and to have a direct relationship with clients led her to quit her job and take the momentous step of setting up her own business at the age of only 25. She rented a workshop on a farm complex in Wimborne, Dorset, and started with commissions that she picked up via the farm shop and ads that she placed in local papers. She held fast to the high standards that had been drilled into her at Sunseeker and charged her clients less than her original estimates if the job took less time. Soon she was joined in the workshop by someone who taught her more of the traditional upholstery techniques in which he was especially skilled.

It wasn't all plain sailing, though. During the early, lean months, Sonnaz did bar work to help make ends meet, and eventually started to do some teaching. The turning point came when a local building firm commissioned her to upholster walls in a client's house – a woman who is still a client today. After that, things started to snowball and all her work has come about through personal recommendations ever since.

Five years later, Sonnaz was contacted by the production team behind *The Repair Shop*, who were just about to launch the first series. At the time she was reluctant to take matters further because she was just starting to get a handle on her business and didn't want to upset her clients. But when the team approached her again, once the show had been recommissioned, and she visited the barn to meet the other experts, she knew she wanted to get involved.

6

The beauty of being dyslexic means that you memorize everything and my mind works in 3D, working out problems in my mind before they happen. Over the years, I've learned a lot about fabric. I love what fabrics can do to a room. Your home is so important to your well-being and as an expression of who you are.

When you're working with fabrics, a good deal of it is trial and error, and, of course, you remember the mistakes the best – the times when you use the wrong weight, or when the fabric drops more than you expected. I once made a pair of curtains that were 7m (23ft) in height, and when they were hung, the fabric dropped by 7.5cm (3in), so I had to hem them on site.

Some upholsterers will only work with foam, some will only do traditional pieces, but I like doing both. With traditional upholstery, it's the springing that makes the seat bouncy. It's stuffed tight and stitched tight and always feels firmer. More modern furniture uses foam, which is very soft and comfortable, although it doesn't last as long – it withers and gets dusty with age.

Working with foam throws up so many challenges. Sizing is difficult to get right, it has to be perfect. You have to judge how it's going to fit into the cover and choose the right density. There are a lot of variables. If I'm reupholstering mid-century modern furniture, the hyper-stretchy fabrics used in that era are no longer available, so you may end up with a join in the top cover in places where it wouldn't have been originally.

When I was working on the King George VI coronation chair, I deliberately used all traditional tools and techniques. I used a tacking hammer rather than a staple gun, and all the materials, the hessian and the webbing were the same as the original. I reused the horsehair pad and put new scrim on top. It's important to honour the original craftsmanship and the integrity of a piece.

For the same chair, we needed some braid for trimming, of the type that isn't available to buy anymore, so I approached Heritage Trimmings who specialize in traditional passementerie. They sent some cotton threads to choose from and I specified a dulled-down version of the original colour so that the braid would look like it had always been there. It's wonderful that such specialist knowledge is still out there. Going the extra mile is really important.

I'm one of those craftspeople who really love their tools. They're as important as your work. The best are the old ones – it's the ergonomics of the way they fit in your hand, the quality of the steel and the metalwork. From my Uncle Frank on my husband's side, I've inherited an old fold-out metal ruler and it's a thing of beauty. The joints and hinges are amazing.

I love a good pair of shears. I've got one pair I use for everything, another which I use for leather because there's a different angle on the blade, and a very fine pair for cutting top fabric. I've also got three different unpickers and two different types of bradawl, so I can choose which to use for which project.

My first Repair Shop job was an iron nursing chair, which was very tricky to do, so much so Jay was impressed that I had the nerve to repair it on TV! But what has really impressed me about the show, ever since my first visit to the barn, is the focus on care and integrity. My career restoring furniture has shown me just how much different pieces mean to people.

Ball Chair

A 1960s design classic

Before Nina Tillett's beloved Uncle Tim died, he and Aunt Janice gifted her their classic 1960s ball chair to honour her own design career, as well as their special bond.

Tim and Janice had reupholstered the chair in the 1980s, and when she was a child, Nina, who lives in Wiltshire, spent a lot of time at her uncle and aunt's house. The chair became a constant in her childhood – a part of the family, as she puts it.

Every child who grows up with a swivel chair knows that even if it's an expensive designer piece, it's also a plaything. Uncle Tim and Aunt Janice never minded how many times Nina swivelled around in it, screaming with delight. The cocoon shape of the chair provided young Nina with a private place to hide away, read and dream.

She was absolutely devastated when her uncle suffered a sudden heart attack in 2012, which turned him into a shadow of his former self. When he died, Nina was bereft. They had an incredibly close bond.

Uncle Tim and Aunt Janice gifted the chair to Nina because they knew how very attached she was to it. And Nina was touched beyond measure, because she knew how much the chair meant to them.

The couple had bought the chair on a whim in the 1970s after falling in love with it in a Manchester shop. They spent their joint monthly wages on it, and both needed a brandy or two before making the purchase.

At the time, they had no other furniture apart from a bed. When it was delivered, it would not go through the door, so they had to have the window taken out to get it into the room.

All these years later, Nina, who is a visual merchandiser, still loves the chair but she hides it away because it's in such a bad state of repair. She still uses it – as her 'thinking chair' – but keeps it downstairs in her workshop, away from her main office upstairs.

Since her Uncle Tim passed away, Nina has found pennies in the back of the chair, which remind her of him. She even discovered one recently in the chair itself and took it as a sign that she should bring the heirloom to The Repair Shop.

Over time and following multiple houses moves, the once smart designer piece looks tired and worn. The original mustard velvet upholstery – like something out of Austin Powers, Nina says – was replaced 40-odd years ago with a 1980s print, which is now faded, and the inner stuffing is worn and squashed. The ball chair looks a long way from how it should.

Repair requires the expert work of upholsterer Sonnaz Nooranvary, and metalworker Dom Chinea. Between them they hope to return the iconic 1960s piece of furniture to its former glory. They are both excited at the prospect of working on such a cool and unusual item.

Sonnaz strips out the upholstery, and cuts and shapes new, thicker foam slices for the back and base. Meanwhile, Dom fashions a fibreglass mould of a section of the outside of the chair for Sonnaz, so she has a solid base when covering the foam. Before fixing layers of fibreglass matting to the outside of the chair, Dom applies a wax mould release, to prevent any damage.

This is an interesting challenge for Sonnaz. She's seen all kinds of furniture, from bespoke yacht interiors to horsehair-stuffed antiques, but never before has she encountered a chair quite like this.

Sonnaz cuts out a pattern for the new covers, then cuts out the fabric pieces and glues them to the foam slices, with the help of Dom's fibreglass mould. Now comes the moment of truth. Will the covers fit?

Together Sonnaz and Dom join forces to reunite the chair with Nina for the big reveal. But not before adding their very own penny down the back of the chair for luck – and old time's sake.

Nina is overjoyed with her newly upholstered chair but it also brings a tear to her eye, as she thinks back to her childhood which she spent swivelling around in it at her uncle and aunt's home.

the
bookbinder

Christopher Shaw

At home near Brackley, a market town in Northamptonshire, bookbinder Chris Shaw works for at least a few hours every day of the year, even on Christmas Day. He has been dedicated to his craft for more than 40 years now, creating and restoring beautifully bound books for Booker Prize winners, famous illustrators including the late Ronald Searle, creator of the original St Trinian's cartoon strip series, and celebrities such as the fashion designer Jean Muir, for whom he produced elegant bindings.

Chris grew up in Basingstoke, Hampshire. There is no history of bookbinding in his family, although Chris still uses some of the tools he inherited from his grandfather, who was a leatherworker.

After leaving school, where he enjoyed woodwork and metalwork but managed to fail quite a lot of his O Levels, he went to sixth-form college to undertake exam resits. It was here that he met his future wife, Sarah.

Christopher is now regarded as one of the leading experts in bookbinding in the UK, and he has travelled all over the world through his work. He became a Fellow of Designer Bookbinders in 2004.

At the age of 18, Chris and Sarah made a pivotal design decision together, to spread their wings and leave Basingstoke for Guildford College of Further and Higher Education, in Surrey, and study together for a diploma in bookbinding and conservation. Chris had always been interested in illustration and was inspired by the work of famous late 19th- and early 20th-century illustrators such as Arthur Rackham.

From Guildford, Chris and Sarah moved to Wootton, near Woodstock, in Oxfordshire, where they set up their own bindery, before settling near Brackley 30 years ago. At first, as they worked hard to establish the business while raising a son and daughter, Chris did night shifts in a plastics factory to pay the bills.

Working in collaboration with artists and printmakers such as Mark Hearld, whose lithographic and linocut prints celebrate the British countryside, Chris is now regarded as one of the leading experts in bookbinding in the UK, and he has travelled all over the world through his work. He became a Fellow of Designer Bookbinders in 2004. One of his favourite jobs is tooling

with gold. He says that any book can be improved with gold tooling, although it can be an 'epic task' for a bookbinder.

Chris joined The Repair Shop in 2020, when he was asked to the barn to work on a 230-year-old dictionary from HMS *Victory*, which had been passed down through the generations to Adam Luxton, who believes from the inscription – 'Nelson's book' – that it was originally owned by Horatio Nelson himself.

The dictionary was left to Adam by his grandfather, Lionel, who had recently passed away. The family story went that it came into the possession of an ancestor who had been a midshipman on the famous warship, which proved such a decisive force in the 1805 Battle of Trafalgar against the combined naval fleets of France and Spain. Following a thorough restoration of this delicate family heirloom, which had been badly repaired in the 1990s, Chris added a thoughtful finishing touch: a protective slipcase with a nameplate bearing Lionel's name. Such sensitive attention to detail has become indicative of Chris's work on the show.

It feels like life's become a lot more hectic since I joined The Repair Shop. I've been so used to working alone and I don't go out much anyway, apart from to Morris dancing, which is one of my interests. Now I've had to learn to communicate my work to others. At school I was very quiet and I've never taught others. It's been a totally new experience for me. I've never worked with people, only my wife or when I did those factory shifts, and I've had to fit into an environment that I've been rather dropped into.

Normally a bookbinder would only associate with other bookbinders. I suppose that all of us in the workshop are very niche – and bookbinding is a very niche niche – but I do like the fact that I'm interacting with other creative people, and I'm always really keen to learn new things from them. I'm very partnered with Louise [Drover, the paper conservator] who does paper restoration. I did a beautiful wedding album with Steve Kember [the music-box expert], and I'm going to be mending a very special tabletop with furniture restorer Will Kirk.

I can totally focus, however, to the point of not really noticing what is going on around me. I become very absorbed in my work. I'm absolutely fascinated by the history of the books I am given to work on. The book itself, the bookplate, a signature, a date or inscription. It's something that you can actually hold, it's a physical item, it's very tactile. All that information places it and brings it alive. That's what the programme does for people, apart from the skill of repairing the object.

Everyone in the workshop is a total nerd like yourself. I love that. I'm in awe of Sparky Mark [Stuckey, the electromechanical engineer] but really struggle to understand his area of expertise. I love to talk to him, and he really dumbs it down for me, but when he starts talking about resistors and currents and so on, it feels like I'm back at school in a really difficult physics lesson! We are collaborating on a lovely project, though; it's a "secret squirrel" radio hidden in a book.

I'm really bad at anything to do with technology. It's just one of those things. I've an inordinate amount of patience with my work, I can sit for hours. But I have very little patience with the internet. Until recently I never even had a mobile phone. And when I did get one, I had to ask my children to sort it out for me.

My grandfather was a leatherworker, with furniture. I have some of his tools. One of the tools is a bone folder. Used for making strong, sharp creases in paper, it's like an extension to your hand.

It must be at least 80 years old. I like the idea of continuity in using it. You really do get a feel for your tools. I've had this particular paste brush for 20 years. I know it's breaking. I'm going through a load at the moment, testing them. There is a going to be a transition soon and I'm preparing myself for it.

I work a lot with Louise, she knows painting restorer Lucia Scalisi well and I think that between them they came up with my name when the producers were looking for someone to restore the Nelson dictionary. First I got a phone call, and then they wanted to see me on a video call. The thing I've always thought of and dreamed of, if I had a proper job, would be to have a hat and my pipe. That gives a certain air of intelligence, I thought.

You know when you see an athlete running? They make it look effortless. That's the aim for me, before I get too old. I never feel that I've done a good enough job. There is so much responsibility in getting it right. I think my favourite project so far was the suffragette book signed by all the Pankhursts. I just kept looking at it, going over all the signatures. I would have love to have kept it, but, of course, it wasn't mine to keep and I was just happy to have worked on it.

But the most memorable Repair Shop project for me has been the prayer book from a concentration camp. When it came in, it was a real struggle to work on it. You realize the enormity of it all; there was a drawing of the concentration camp gates on the front, a poem, and 50 survivors had put their signatures in it. I was terrified of what I was saying, I was very aware of the respect needed and didn't want to belittle it. I've enjoyed them all, but this was the most important book I've ever restored. I never thought I would get emotional about an object; it was unbelievable. The Repair Shop does that, it makes these cherished family items so much more visceral.

Sometimes when we go for a reveal, there is nervous excitement, but on that one we were all so reverential. It was almost silent in the workshop. You could hear a pin drop. We were all either in tears or close to tears. It's moments such as these which remind you just how much the work you do means to people. The grandson of the book's owner, Gary, sent me an absolutely beautiful email afterward. I was really taken aback by that. He was so kind and appreciative. You don't realize what you're doing for people, how you're mending and saving their memories and family history, and that means such a lot to them.

Jewish Prayer Book from a Concentration Camp

A World War II survival story

Sometimes an object comes into the Repair Shop and it is so significant that everyone simply stops what they are doing and takes a moment to appreciate how privileged they are.

When Gary Fisher, a retired senior bank manager from the West Midlands, arrives with a Jewish prayer book that belonged to his grandparents, Emanuel and Gisela Fischer, this is what happens. The prayer book is a Haggadah, read by Jewish families at the beginning of Passover. It is small, but the life experience embodied on its pages vast, fathomless even, spanning centuries and countries, war and peace.

Not only is this precious book almost 100 years old, it has survived the horrors of Theresienstadt Ghetto, a World War II Nazi concentration camp in the Czechoslovakia. Gary's grandparents were imprisoned here in 1942. He tells us that they were only permitted to live as Gisela, a trained nurse, was deemed 'useful' by their captors. Gisela and Emanuel lost many family members at the hands of the Nazis.

This prayer book provided the couple with hope during one of history's darkest moments. They were allowed to practice their religion in the camp, Gary says, because the Nazis used Theresienstadt to hide the fact that they were murdering people in other camps.

In the book, his grandfather drew the Star of David in the gates of Theresienstadt, and wrote a poem, in Hebrew, about his imprisonment. After their liberation, Emanuel and Gisela collected the signatures of about 50 fellow survivors in the book, so that they would never be forgotten. Their 16-year-old son, Harry – who was Gary's father – had been one of the 10,000 Jewish children taken on the Kindertransport from their home in Austria to England in 1938. Harry played his own important role in history. As a young man, he became an interpreter and after the war, even attended the Nuremberg Trials, in Germany, when Nazi leaders were prosecuted for the horrors of the Holocaust.

In 1943, Harry married Gary's mum, Joyce, a sergeant major in the Women's Royal Army Pay Corps (WRAPC). Harry thought his parents were dead, until after the war he received a telegram telling him that they were alive and in Vienna, Austria. They then joined their son in England.

Gary's grandparents and their prayer book played a significant role in his life growing up. Every summer, his parents would put him on the train from Leamington Spa, where they lived, to London to see them; his memory is of 'the kindest people you could ever think of'. Gary's

grandparents would read passages from the prayer book in Hebrew.

The prayer book helped Gary's grandparents when all seemed lost, and they never forgot its importance, treasuring it and using it every day. Now, understandably, it is showing signs of age. However, Gary's great-niece, Heidi Owen, is now at university and writing a book about the life of Gary's dad and grandparents. It is this interest from the next generation that has inspired Gary to approach The Repair Shop and ask if anything can be done to help return this precious tome – which has survived so much – to its original condition. The prayer book is falling apart, with missing binding, the cover worn and the pages torn.

When Gary reads out loud the poem his grandfather wrote – he has had it translated into English from the original Hebrew – bookbinding expert Chris Shaw bows his head and is visibly moved. As Chris examines the prayer book's fragile state, he explains that it has been made of sections stapled together. The metal in the staples has reacted with the paper over time, creating rust-coloured marks.

Chris decides that the best solution is to remove the staples and replace them with linen stitching to hold the book together. He also wants to repair the book's edges. The first thing he does is dismantle the whole thing and separate it out into sections.

He admits to being nervous; he does not want to disturb those precious signatures. So, he uses his scalpel to carefully peel back the board surface in order to introduce delicate sheets of new paper underneath to stabilise the pages. 'It's just so epic', he murmurs.

When the book is stable again, he can turn his attention to the restoring the damaged pages and the cover. He uses a soft black watercolour to 'dob' in a few of the chips on the cover. However, he leaves some of the rubs and marks intact, so as to preserve its history for future generations.

When Gary returns to the workshop, he says he feels his grandparents are with him, waiting to see the result. Chris looks nervous, but there is no need for his concern. Gary is thrilled. 'It's a complete work of art', he says. Chris says it has been the most important book he will ever repair. He couldn't let it go without 'extra protection', he adds, so he has also made a slip case, engraved with the names of Emanuel and Gisela Fischer as an everlasting memorial.

The history and significance of the prayer book made it the most important repair Chris had ever undertaken. He was committed to doing the best possible work he could for Gary in honour of his grandparents who, with their precious book, had survived a Nazi concentration camp.

the
cobbler

Dean Westmoreland

When cobbler Dean Westmoreland was still at school in Bradford, West Yorkshire, he had a part-time job at Wickes, the building supplies merchants, because his dad worked in the building trade, along with some of his uncles and cousins.

He could have easily followed in their footsteps, but Dean had another ambition: to be a rock star. He played guitar and took his first job only because it helped to pay for guitar strings and rehearsal sessions.

After a couple of years selling bricks and blocks, he moved on to another part-time job, this time in JJB Sports, a sports store in Bradford, fitting trainers. He had always loved footwear, and with his very first wage, had bought himself a pair of Clarks shoes. Although he was still chasing his dream of rock stardom, it was Dean's experience of fitting trainers that led him directly to the door of The Repair Shop.

... the smell of fresh leather and the challenge of making battered boots look as good as new again began to excite him even more than being on stage blasting out the chords.

Spotting an advert for what he calls 'a glorified tea boy' role at the Bradford branch of Timpson, the well-known chain store of cobblers, he found that the smell of fresh leather and the challenge of making battered boots look as good as new again began to excite him even more than being on stage blasting out the chords.

He was fortunate enough then to be able to train in the craft of cobbling with master repairers. Over an apprenticeship lasting four years, Dean was taught intricate techniques such as pinwork, working with masters of the craft, including his mentors Rachel Matthews and David Balderstone, the veteran cobbler who passed away in 2012.

However, in 2010, Dean went part-time and returned to Leeds College of Music to finish off his degree in music. When his fiancée became pregnant, he finally admitted to himself that he wasn't going to be a rock star – although he does still practise his guitar every week – and decided to concentrate on honing his cobbling craft.

A series of jobs working with independent cobblers in West Yorkshire gave Dean the experience to enter international cobbling competitions. In 2016, he was named the J. Rendenbach Shoe Repairer of the Year – the last cobbler to win this prestigious industry award. It's one of Dean's ambitions to find a sponsor for a new award to honour the very best practitioners of his trade, as the J. Rendenbach tannery in Trier, Germany, closed in 2021.

In 2017, Dean opened his own cobbling business, Yorkshire Sole in Shipley, West Yorkshire, specializing in the repair and restoration of prestige brands such as Crockett & Jones, Tricker's, Loake, Cheaney, Church, Oliver Sweeney, Barker, Jeffery-West, Sanders and Red Wing boots – his ticket to *Repair Shop* fame.

In May 2020, fellow resident expert Dom Chinea posted his own battered pair of Red Wings on Instagram and asked for ideas for what to do with them. Dean sent Dom a direct message telling him he could help bring his treasured boots back to life. A couple of weeks later, an email arrived from *The Repair Shop* producer, inviting Dean to get in touch.

At first he thought it was a prank, but after being invited down to the barn in Chichester, he realized that the invitation was serious. Now, at 33, Dean has become one of the latest experts to join the show.

There's a perception of a cobbler as a cranky old man working alone in his little workshop – and I'm the opposite of that. I think that the producers took to me because I'm youngish for the trade. I hope that people see me as honest, proper Northern and that I care.

At the beginning, after a couple of shows, one of the producers said to me that I really listen to the contributors. I'm really empathetic, apparently. I don't know about that, but I do feel what the guests feel; nervous, excited, hopeful. It can be very emotional.

You really do feel all of that very intensely at the reveal table. Are they going to like it? Have I done a good enough job? That all goes through your mind.

I feel very privileged to be part of The Repair Shop. There are other guys doing such great work, and I still have to check myself sometimes when I realize that I'm fortunate enough to be part of that team. I'm a relative newcomer really and it was quite difficult at the start. These people are almost superstars in their own right, but they were all so friendly and really welcoming

It's very rare that you would be there in a group with so many different experts in their field, but that's what The Repair Shop does, it brings everyone together. The mind-set is very similar. There's a lot of integrity in what all the craftspeople do. The atmosphere is relaxed, but quite intense. Everybody is trying to excel in their craft.

Doing this has given me a lot of confidence. I've worked on nine or ten projects now, with some collaborations. I've pulled in Suzie [Fletcher, the resident leatherworker] on a couple of tricky stitching bits. I like to flesh out ideas with her. And I've worked with Dom [Chinea, the metalworker]. A lot of things that I mend have metal on them. I have so much respect for what they do and value their opinions.

The first project I worked on was a pair of running spikes that sprinter Audrey Brown wore when she won a silver medal in the 4 x 100m relay at the 1936 Berlin Olympics, which were watched by Adolf Hitler.

These spikes were brought in by Audrey's grandson Tom Wenham, from Ilkley in West Yorkshire – he's a coach for Team GB men's lacrosse. Audrey had died in 2005, aged 92, leaving Tom the bespoke leather spikes wrapped in a tea towel.

I was really nervous. I thought to myself, I can't believe it, for my first project they've given me something to do that I've never done before: running spikes! But when I got the item, I could see the way they had been constructed – an ancient way of construction that nobody does any more. I was pleased that I managed to figure out a way to fix them.

This project obviously stands out in my mind because it was the first, but one of the most moving and memorable was the speedway boots I mended. The gentleman they belonged to passed away just as we were finishing the repair; the reaction of his wife and daughters was so emotional.

I am quite an emotional person, I must admit. I've got a lot of love for my knives – we're very attached. I'd say they are definitely my favourite tools. They're shoemakers' knives; a lot are curved, with a bevel on one side. When you're working with it, the bevel naturally takes the knife away from the shoe. The knives have different curves for different shapes. A lot have been gifted by other shoemakers; a lot I've picked up off eBay. I get my new knives from Germany. This is a shame as they should be from Sheffield, but German knives keep their edge. It's taken me about three years to work out the best way to sharpen my knives. You don't want to be sharpening after every cut; you just want to run it once on the strop and get on with the job.

My favourite shoes to work on are Goodyear Welted shoes. These are proper shoes made in the traditional way, and it's the highest spec of shoe construction. I love the traditional projects. There's a pair of officer's all-leather dress wellies coming in soon and I'm really looking forward to working on them.

I'm hoping that I can repair more shoes and give more of a spotlight on my trade as a whole so that it's not forgotten – it should be celebrated. Although we cobblers use very old techniques, repairing footwear is having a renaissance because people are buying things to last more these days. Also, repairing is very much in tune with the upcycling trend. For myself, I just want to be the best I can be. I thought I'd be playing Wembley Stadium and here I am, in The Repair Shop. And I wouldn't change a thing.

Roller Skates

Rolling back the years

A pair of white leather roller skates, more than 70 years old, stand as a touching reminder of a shared hobby which sparked a romance between two much-loved parents. Now the children and grandchildren of Irene and Colin Tingle, who met at their local roller-skating rink near Barnsley in South Yorkshire in 1949, would love to see the wooden wheels of these treasured skates turning once more, especially as they could be a perfect fit for one of Irene and Colin's great-grandchildren.

Irene died at the end of 2019. Eve's mother, Angela, rediscovered the roller skates when she was sorting through her things, plus a skating dress and newspaper clippings, including a report of her parents' 1952 wedding from a local newspaper entitled 'Roller Rink Romance'.

While Angela and her sister admit that they didn't inherit their parent's excellent sense of balance, they would love the skates to be enjoyed again and not just kept in a cupboard, never seeing the light of day.

However, the once-pristine white leather is scuffed, dirty and damaged. The metal components have rusted and multiple metal eyelets are missing. Eve, Angela and the rest of the family would really like them to be brought back to live another lifetime, and so they can show Colin, who has had Covid twice and is now 90 years old.

Angela says her parents had a very close bond, even though they were quite different; her mum took great pride in her appearance and always had her hair and nails done. She loved to travel, while her dad was quieter and liked to stay at home. Her dad was heartbroken when her mum died and the family have had a difficult time helping him to come to terms with his loss. He is still missing Irene dreadfully and so is Angela. She feels that the skates might remind her dad of those carefree days when they met and the happy times they spent together.

Eve loves the fact that the roller skates are what brought her grandparents together and were such a special part of their life and their family history.

Cobbler Dean Westmoreland is itching to get his hands on these 'beautifully made' boots and have them rink-ready again.

Dean admits he has never tackled a pair of roller skates before. The first thing he notices is that one of the boots has a tear on the toe. Next, he picks up on the many missing eyelets, and the fact that the battered white leather needs cleaning and smartening up.

To fix the tear, Dean adds a membrane patch underneath the affected area. When the patch has set, he covers it over with fresh white calf leather. The next stage is to sort out the eyelets – 40 in total need removing and replacing. He has sourced some new white replacements, which will blend into the boot once the leather is dyed with acrylic dye.

The first coat of dye is always slightly patchy, he says, but the second coat will bring the creamy white leather to life. Dean carefully fixes the wheel mechanism back onto each boot when the dye has dried thoroughly.

When Angela and Eve return to the barn, Eve admits she feels a mixture of emotions; excited, nervous, anxious. An emotional Angela says that her mum is on her mind on a daily basis. She's thinking of her dad too and how he will remember her mum wearing the boots.

However, mother and daughter bring with them exciting news; Eve is expecting a baby, and Angela says that she hopes that the skates might eventually fit one of her grandchildren.

She's overcome with emotion when she sees the skates gleaming white once more, just as they were in the photographs of her beloved mum when she was young.

Dean hopes that Irene's restored skates will help to lift Colin's spirits as they remind him of their fun roller-skating years together.

the organ builder

David Burville

As a child, David Burville would travel the country with his father almost every weekend, appearing at steam rallies, fairgrounds and charity events with a huge, brightly painted Belgian dance organ the size of a double-decker bus. This colossal instrument, pulled on a trailer, had been inherited by his dad, a college laboratory technician, from a family friend who had brought it over from Holland in the 1960s.

David's great-grandfather, Harry, was well connected in the world of showmen, and David's Dad helped out at fairgrounds in his youth, putting music through the fairground organs and running the coconut shy.

His love of fun and entertainment had passed down through the family, and this is why David's dad was chosen to become the custodian of this beautifully carved and gilded instrument. It was to prove pivotal, however, not only by shaping David's unusual 1980s childhood – through this organ, he was introduced to famous people at events, including the Prince and Princess of Wales – but in laying the foundations of his future career as a specialist organ restorer.

David is known for his multitude of
skills, which embrace cabinetmaking
and furniture restoration, as well as
the intricate mechanics of musical
instruments.

When he left school, David, who grew up in the village
of Preston, near Canterbury in Kent, decided to embark
on an apprenticeship with a local company that built
church organs.

Although he enjoyed learning about the mechanisms
of church organs, from building through to completion –
as well as visiting the beautiful churches and cathedrals
that came as part of the job – after a while David grew a
little restless. He missed the vibrant fairground organs
of his childhood. So, when he was 18, he took up the
invitation of an apprenticeship with J. Verbeeck in
Antwerp, Belgium, a leading organ manufacturer and one
of the largest and most influential in the world.

After two very happy years with the company,
learning everything he could about fairground and
mechanical organs, David returned to England to settle
down. Following a three-year stint back with the church
organ builders, David made the decision to become self-
employed and set up his own organ restoration business
in Canterbury, Kent.

When The Repair Shop found itself with a harmonium
in need of some serious TLC, out went a call to David's
workshop. With his positive can-do attitude, he didn't
hesitate to respond and after a short video call, he was on
his way to the barn to lend a hand.

Now David is known for his multitude of skills, which
embrace cabinetmaking and furniture restoration, as well
as the intricate mechanics of musical instruments. He still
repairs fairground organs for his own business, but on
the show his expertise in metal engineering, mechanisms,
crankshafts, flywheels and, of course, carved, painted and
beautifully gilded woodwork, is always in demand.

Generally, I'm a very positive person and always smiling and laughing. I can get on with anybody, and the thing with The Repair Shop is that everybody is like-minded. We're busy all the time, but there are lots of people to have a laugh and a joke with. Musical instrument specialist Pete Woods, 'guitar doctor' Julyan Wallis, Mark the Spark (aka radio expert Mark Stuckey), we all get on very well together. We're always larking about. Sometimes when things are getting intense, I think we bring a bit of energy and light-heartedness to the barn.

One of the most memorable moments for me has been a model fire engine. My grandad, my dad's dad, was an auxiliary fireman in Canterbury during World War II. In their downtime, the firemen used to make toys for the children at Christmas. Everyone would pitch in, according to their skill. Someone good at woodwork would do all the wood, the metalworkers would do the metal elements and so on. It was rather like The Repair Shop in that respect.

This model fire engine, which was brought in by the grandchildren of the chap who had made it, was very poignant to me because of the links to my own family's past.

It was a 1930s-style fire engine with a big ladder on the back, a beautiful model but very decrepit now. It was the most fantastic project - right up my street -and needed a lot of work to get it up to scratch. Everything had to be re-painted and there were little parts missing.

The fire engine had wiring for the headlights to work, driven by a motor, but the headlights had been removed. Because it had been made in the war, it had been built out of anything the chap had been able to get his hands on. A lot of it, including the gearbox and mechanism, had been constructed of Meccano.

There was only one light working, the searchlight, When I looked at it, I thought, "I know what that is". It was the brass part of a bayonet light fitting, so I went into B&Q and bought some and remade the headlights so they would work again.

I absolutely love that flexibility and the collaboration aspect of what we do. It allows us to take on projects that one person wouldn't necessarily be able to do on their own. And this means that we can help more people. I've been with The Repair Shop for just over three years now and I've done collabs with all sorts of people; The Teddy Bear Ladies, Dom Chinea, Steve Kember, Steve Fletcher.

When you're collaborating, you're aware of the significance of these people, very much in awe of what they're doing. You almost don't feel worthy. But we love bouncing things off each other, it's reciprocated backward and forward.

For me, dealing with the emotions of it all is the big challenge. You certainly hear the stories surrounding the objects and you do naturally get a feel for exactly what these precious things mean to people. I tend to get more emotional when an object was physically made especially for a person, such as the model boat, a radio-controlled battleship, a girl brought in that had been made for her grandfather when he was a boy.

I do have a strong emotional side. There have been times when I've been standing at the reveal, when the anticipation of what the guest is going to feel and how they're going to react when the cloth comes off is really intense. I've lost count of the number of times when I've had tears flooding down my face, a mixture of relief and pent-up stress.

One of the most emotional ones was when a gentleman brought in a tabletop Indian harmonium. When we had mended it, he actually sang a traditional song and wanted us to join in. To see him getting so much pleasure from what we had done, the absolute delight and emotion on his face, was amazing.

So, yes, when that cloth comes off, I'm as nervous as hell, because you've been getting to know that item, learning about its background and what it means to the person who has brought it in. You cannot mess it up. I've worked on a few toys and to get something wrong would make that item alien to those who loved it. I worked on a little giraffe, a string-and-bead toy. His eyes had gone missing and, at first, I couldn't quite see how they should have been. The eyes are the most important thing; to get them wrong would have meant that he wouldn't have looked like the toy his owners remembered. I did get it right in the end, I hope.

Over the last couple of years my work here has really influenced my life. I think I've spent more time at The Repair Shop than in my own workshop. My favourite tool is probably my miniature turning lathe. You wouldn't believe the amount of different things you can do on it, and it allows me to make some quite complex bits and pieces.

I don't think that any of us in the barn would be comfortable with the term "expert". Yes, you've got knowledge in your own field, but we're all learning all the time. We're willing to accept that we don't know it all and to ask the others to pitch in. There is always someone who is willing to help out and lend a hand so that we can achieve our aims.

Barrel Organ

Getting back in tune

Our super-enthusiastic organ expert David Burville can hardly contain himself when Norman Kench and his daughter Sarah, from Warwickshire, arrive at The Repair Shop. They've brought with them a late Victorian/early Edwardian barrel organ which has seen better days and are desperate for him to take a look.

It plays 10 tunes, Norman jokes, 'badly'. And he's right. The poor old street instrument makes quite a racket. He thinks it's been in Henley in Arden for at least 100 years, and now it needs 'a bit of TLC'.

Norman is curator of the Henley in Arden Heritage Centre. His wife, Pat, who died four years ago, was a committed local volunteer too and would have been delighted to hear the barrel organ working properly again. It belongs to the town, Norman says; the plan is to find a home for the instrument inside the Heritage Centre so young people can see and hear what their great-grandparents would have been entertained with.

Since the Heritage Centre opened, the barrel organ has been a regular town favourite, appearing at events including Christmas on the High Street. Its 10 songs include *My Old Man Said Follow The Van*, *A Broken Doll* and *The Honeysuckle and The Bee*, but Sarah laughs that at the moment they all sound the same.

David points out that the barrel organ started life as a 'café piano' which would have had a slot for coins, allowing customers to choose their favourite tunes. A lot – like this one – were converted to run on carts, and used by World War I soldiers, he adds.

The instrument has been 'bashed about a bit', and the veneer is lifting in places. But the most important aspect is the mechanical condition. The pins of the barrel operate onto hammers, which strike strings and produce notes. On the face of it, apart from being dirty, David says the interior doesn't look 'too bad'.

He takes it to pieces to see what's wrong with it. The lower notes are made with felted hammers to give them a softer, warmer tone, he explains, and the plain, wooden hammers in the upper melody sections give more of a 'brilliant' hit. The two combine to create that distinctive jingle-jangle sound.

The felting is in reasonable condition, but David finds that, at some point, someone has put leather to the surface, perhaps to alter the tone. He removes all the hammers to take the leather off each one, so he can start to re-dress them all together.

Now the leather is removed, he addresses the felt, which he finds rather soft. So, he uses a very weak solution of shellac and a pipette to put a little drop onto the felt just behind the 'hitting' surface.

This will soak into the felt and make it a little bit harder, giving it a much nicer, clearer sound when it is

struck by the hammer. He must repeat this process for all 47 hammers, then return them to their frame and fit the frame back into the instrument.

The barrel, David says, is in good condition. He returns this and the frame to the casing and prepares to tune it all up with an electronic tuning meter. He sets the tone to 'A', takes the A string by hand and plucks it, knocking it gently around until it sounds really good. There are hundreds of strings to go through, but David says it's really worth the time and effort.

Now it's time to roll out the barrel again. David and Jay wheel it back into the barn and prepare to face the music - Norman and Sarah. Norman says a space has been cleared in the Heritage Centre ready for the return of the town's favourite.

'It's amazing,' he says. David has given the casing a polish and it gleams, but Norman is most interested in hearing how the instrument sounds. David turns the crank handle, a jolly tune plays and everyone breaks out in a round of applause before Norman has a practice. It's going to be his job to turn the handle from now on.

'I didn't think I'd ever hear it go properly again,' he says. He and Sarah agreed that Pat would have been so pleased and proud.

As an organ restorer, David Burville was delighted to get his hands on this original barrel organ.

the bicycle restorer

Tim Gunn

Tim Gunn grew up in a world of transport enthusiasts. His father restored vintage cars and his grandfather had been a keen cyclist. It was when his father took his grandfather's bike, a 1934 Sunbeam, to have gears fitted that a seed was planted in Tim's mind. The man who supplied and fitted the gears was a collector of old bicycles – everything from penny farthings to Victorian boneshakers – and it seemed to Tim that he had arrived in bicycle mecca.

During his teenage years, Tim was always tinkering about with bicycles, going soapbox racing and building go-karts. At a local fair, he bought a 1930 BSA roadster, which proved to need a lot of work doing to it. Subsequently, he saved up for a penny farthing, taught himself to ride it, and recreated a journey that his grandfather had once made, cycling from Essex to Gloucestershire over a period of three days.

Meanwhile, he had gone to work at a fish farm, where he was in charge of looking after the koi carp that the company imported from Japan. However, just after his 21st birthday, he decided to quit his job. At that point it occurred to him that, unlike vintage cars, no one specialized in restoring vintage bicycles.

The Repair Shop came into the picture, asking Tim to repair a penny farthing. It was the first of many restorations he would tackle for the programme, and almost a return to where he had started off in the bicycle business.

From a modest beginning, buying, selling and restoring old bikes, Tim's business began to take off. An article published in a French car magazine, written by a journalist friend of the family, brought a stream of French clients to Tim's workshop – English cars and bicycles were highly sought-after in France.

As he delved deeper into the market, Tim discovered that most bicycle manufacturers were moving over to mountain bikes and virtually abandoning their previous models. This gave him the idea to buy up the old stock and sell the parts on his website. Soon he had a huge mail-order business with customers from all over the world. In due course, he became the second largest UK retailer of original Brooks leather saddles, as well as selling whole bikes repaired with parts he held in stock.

But eventually the bubble burst. Brooks decided to sell direct to the public, and Google changed its algorithms so that visits to Tim's website nosedived, with disastrous financial consequences for his business.

What rescued Tim's fortunes was the opportunity to work on a project for Pashley, the last traditional cycle manufacturer in the UK. He was put in charge of Pashley's brand of GB cycle components. Several years later, when that came to an end, he went to work for someone who rebuilt vintage Rolls Royces.

At that point, *The Repair Shop* came into the picture, asking Tim to repair a penny farthing. It was the first of many restorations he would tackle for the programme, and almost a return to where he had started off in the bicycle business.

When a bike comes into the barn, it's generally worn out, broken and missing parts. It might be seized up with rust after being left outside for long periods of time. Rust is tedious and time-consuming to fix, but the chrome work is generally recoverable if the rust is not too deep.

My previous experience means that it's simple for me to source parts that are missing or broken. What is harder is restoring a bike so that it still looks the way the owner remembers it. Brand new components look out of place.

The penny farthing, which was my first job at The Repair Shop, was missing the saddle spring, which acts as the suspension. It would have been easy to take a piece of metal and bend it in a vice, to do the job. Instead, I formed it all in the correct way, and used the forge to stamp out the hole to go over the headstock of the bike. I shaped it, using the type of file that would have been used at the time and attached the new seat spring to what remained of the old. For me, it's important to use as much as you can of the original. If you're working on a bike that dates from the 1880s, you've got to bear in mind what tools, materials and facilities were available at the time.

I like using old tools. They give you a connection with the past. I've inherited a lot of my grandfather's and father's tools, and acquired others over the years. Many of the tools that were in cycle shops in the early 1900s, like wheel-building jigs, were made of proper cast iron. I've got a shear tool for cutting spokes with a gauge on one side that tells you how long the spokes should be, and another tool that rolls the thread to the spoke. They're over 100 years old and they still work. It makes you wonder who bought them, who made them and whose workshop they came from.

I do have new tools, such as a modern bench grinder. I also have a power file, but I don't use that very often because it is easy to take too much off. If you use a hand file, you get a feeling for the metal you are trying to manipulate and can see it happening in front of you.

Another process I use is brazing, which is a form of brass welding. All early steel bicycle frames were built using that technology. Once you heat the brass with a flux, it flows like water within the joint, whereas with standard welding, you're only covering the joint, which is not as strong.

When I first turned up at the barn to repair the penny farthing, I was at a pretty low ebb. Losing my business had given me stress alopecia, and without my hair I felt like I had lost my identity. The repair itself was quite a struggle and took me ten days. But putting the soul back into that bicycle gave me back my mojo. The Repair Shop kind of repaired me.

Child's Tricycle

Keeping the wheels turning

Charlotte Lewis remembers the awful feeling of being left out of 1990s childhood summer adventures as all her friends sped off on their bikes to the park, leaving her behind. Small for her age and with little legs, she could never get the hang of riding a two-wheeled bicycle.

Then Charlotte spotted a sturdy vintage tricycle in a neighbour's garden, and her mum agreed that a three-wheeled alternative would be the answer. Even better, the neighbour said that Charlotte could have it for free.

Charlotte recalls that when, as a five-year-old, she sat on this unexpected gift, she was thrilled that she was able to ride independently and would fill up the boot compartment with her bits and bobs.

When Charlotte left home, the tricycle ended up languishing in Stephney's shed.

Now, with the fifth birthday of Charlotte's own daughter, Cleo, approaching, Stephney would love to see the beloved tricycle restored for her. She has the same diminutive stature as her mum, and also struggles to master a 'normal' bike.

However, it's in a sorry state. The wheels have buckled, the spokes are bent. Even the rubber on the tyres has worn away. Water damage from the shed where the tricycle was stored has caused extensive rust, the seat and handlebars need replacing, while a logo on the frame also needs restoring.

With bicycle expert Tim Gunn leading the tricycle fix, restoration genius Dom Chinea lends a helping hand with the rust and paintwork. First, Tim takes the bike apart to free up the buckled wheels and rusted seat. Next, he hammers out the dents in the mudguard on an anvil. Dom collects the frame for de-rusting and sandblasting. Then Tim uses an interesting technique to straighten a bent wheel; he jumps up and down on it.

With the metalwork gleaming once more, Dom is so excited that he's got the exact shade of baby blue paint to match the original colour, it's quite bold alongside the original deep cherry red, he says. Tim is carefully lacing the spokes back into the wheels; it looks very complicated, but he knows what he is doing.

When the parts come back from Dom's respray and the wheels are finished, it's then back over to Tim to reassemble the tricycle – a very fiddly job, involving endless screws. Stephney, Charlotte, Cleo and her younger brother, Caleb, return, eager to see the transformation of this beloved family tricycle. Charlotte says seeing it looking as good as when she first saw it has taken her right back to her childhood. A delighted Cleo does a lap of honour around the workshop while everyone cheers, and rides away with her family following happily behind.

Being able to ride the restored trike, inherited from her mum, Charlotte, means that Cleo's wish to be more independent has come true.

the electromechanical engineer

Mark Stuckey

From tape recorders to vintage radios, from televisions to telescopes, there are few mechanical, electric or electronic items that Mark Stuckey hasn't worked on or repaired. In a very varied career, which has seen him employed on the research, development and production of classified weapons systems, along with a few years in film finance, one constant has been his passion for learning. His all-consuming desire to find out how things work he traces back to a childhood fascination with the fictional time-traveller Dr Who and his seemingly boundless knowledge. Small wonder that today Mark is The Repair Shop's mechanical whizz.

Mark's father was an engineer who worked for Marconi Instruments, and the family moved frequently from one location to another. Consequently, schooling was disjointed and Mark was keen to leave formal education as soon as he could. His first job, which his father arranged for him, was in mechanical engineering. Hating every minute of it, he quickly took matters into his own hands.

A school friend was working at a local shop that specialized in repairing televisions and radios. Every evening, after his normal work finished, Mark would take himself along there, staying until the shop closed at 10pm. After six months or so, he was offered a job. During his time there he was allowed to go on day release so that he could complete a City and Guilds course in TV and radio servicing.

... there are few left today with Mark's degree of knowledge and experience, as viewers of *The Repair Shop* will appreciate, along with those grateful owners whose treasures he has brought back to glorious working life.

But by the late 1970s, television and radio technology had moved on, and so did Mark. After a year or two working for a firm that repaired photocopiers, cash registers and calculators, an opportunity arose for him to join a company that dealt with Ministry of Defence projects. Here he progressed from the service department to research and development, taking time out to complete a degree in physics. Eventually he ended up working on the design of various different weapons systems, which, he freely admits, involved a high level of skulduggery.

When competitive tendering was brought into the public sector during the 1980s and those in the industry found themselves undercut by commercial rivals, often with less experience and knowledge, it was time for Mark to make another career shift. This time, he took the surprising step of going into the film business, working at Pinewood Studios on the financing of pop promos and television programmes such as *Red Dwarf*.

Then one weekend he visited a friend in Norfolk who had moved there some years before and been extolling its virtues ever since. On the following Monday, as soon as Mark came home, the family home went on the market.

Once he and his family had settled in Cromer, Mark set up a business in workshops in his back garden, repairing radios, televisions, tape recorders, crystal sets – anything vintage that contained electronics. His expertise meant he could repair and restore their cases as well as their working parts. Demand was high. Unlike in the past, when an apprentice could always ask a more seasoned expert for help with a tricky problem, there are few left today with Mark's degree of knowledge and experience, as viewers of *The Repair Shop* will appreciate, along with those grateful owners whose treasures he has brought back to glorious working life.

When an electronic item comes in and it's not working, it's because something has failed. I literally see it as an Agatha Christie, a whodunnit. Someone's done a murder, none of the suspects want to talk, so you have to investigate. In this case my interviewer is my test equipment: my multimeter, my voltmeter, my oscilloscope, my power source. You try to eliminate each suspect one by one, find the smoking gun and the culprit, and make good the repair.

Nowadays, we're looking at items that have gone way beyond their normal expectation of service, which means you're dealing with old-age issues. On top of the repair, you also have to go one step further and replace parts that will fail sooner rather than later, simply because of how old they are. The item may function as soon as you turn it on, but you want to make sure that it keeps on working once it has left the barn and for a long while afterward.

Electronic items have their own personalities and characters. They're sensitive little souls, especially if they're not stored in an ideal environment. When they stop working, people often put them away in a garage, shed or loft, places that have extreme variations of temperature and humidity. Many of these items contain very fine coils of copper wire, and if copper gets damp, the circuit will go.

Cold is also harmful. Cold upsets electronic equipment, especially radiograms. The motor is connected to an idle wheel made of rubber, which turns the turntable or platter. If the idle wheel gets cold, it goes hard and it won't work.

Dust is a different matter. If I open up a set of any kind and it's full of dust, that's a good sign. It means it hasn't been tinkered with. Man-made faults are the hardest to find and repair.

Many types of vintage electronic equipment were very well built and have a sound quality that purists or audiophiles prefer, especially if they contain valves. If a valve hasn't been used, it's as good as the day it was manufactured. Valves are very resilient – unlike a transistor, they can take a lot of hammering and beating. Although valves stopped being manufactured once transistors came in, there is still a lot of new old stock out there. Someone I know who lives less than a mile from me stocks 50,000 of them and if I want a particular valve, he'll have it. Today, due to demand, they're also being remanufactured.

Many pre-war items have cases made of Bakelite. One of the radios that came into the barn had been shipped over from Pakistan by its owner. Unfortunately, it had been smashed in transit and the case was in 30 or 40 pieces when it arrived. I rebuilt it and resprayed it until it looked as good as new. But the glass was also broken, and to replace that we had to find a substitute radio to serve as a donor, which we eventually tracked down in the US. Unfortunately, it didn't make it past US Customs – there's a small percentage of asbestos in Bakelite and it was perceived to be a health hazard. In reality, Bakelite poses no real threat to human health unless you sand it down heavily.

There's a very fine line between repairing and restoring. Sometimes knocks and bumps are all part of the object's history and you want to hold on to that if you possibly can. In the same way, many of the early transistor radios, although they were new technology at the time, were pretty terrible. We're not here to make something better than it was originally designed to be – otherwise you might as well get a brand new radio.

When an item comes into the barn, I try to date it for the owners. Many of the devices have date-stamped components, which means you can sometimes tell within a month when the equipment was originally produced. Then it's possible to give an impression of what it might have cost at the time of purchase. A radiogram I repaired recently, for example, dated from around 1959–60 and would have cost a family 75 guineas, or £3,500 in today's money. That's why they were looked after. But you can't put a price on what they mean to people. You're not just looking at this inert object, you're looking at something that has a real emotional connection to their lives.

When I had young children, I couldn't afford to put my car into the garage for repairs. I had to service and fix it myself. Today, the mentality is different, technology has moved on, and it's harder to repair things. But the greenest thing you can do is to make do and mend, and keep your old things going.

The Repair Shop is very good at shifting you out of your comfort zone. My first repair was a transistor radio. The Dunkirk radio was the second. Then I was asked to fix a jukebox, which I had never done before. It's a fantastic challenge to work on something you have never come across in your everyday job.

Windrush Radiogram

Reviving a mid-century icon

Sisters Fredricka and Louisa Charles would love to hear records being played again on the radiogram belonging to their parents James and Lucy, and remember the happy times at home in the 1960s with a wonderful network of uncles and aunties dancing, singing and enjoying traditional Caribbean food and drink.

Charles and Lucy enjoyed jazz and calypso tunes, especially on Sundays, when their house became a social hub. The radiogram helped to bring everyone together. James and Lucy were part of the Windrush Generation, coming over to the UK from Dominica, an island in the Caribbean. The radiogram, with its smart cream and red cabinet with gold trim and slim mid-century legs, was one of the first pieces of furniture they purchased for their new London home.

After Lucy and then James passed away, the family would still play their dad's favourite tunes on his cherished radiogram whenever they got together. His favourite artists included jazz singer Ella Fitzgerald and vocalist and pianist Nat King Cole. However, it abruptly stopped working about three years ago. They would love it to be repaired to continue to play the music their parents loved.

If it can be fixed, Fredricka plans to put the radiogram in pride of place in her own home and invite all her siblings and their children over to listen to the music that characterized her childhood. Both sisters agree that it would be lovely to remember their parents in this way.

Radio and audio mastermind Mark notices straight away that lots of the components have aged. He wants to strip everything out to see what changes need to be made. Also, the cabinet's good looks are fading. The cloth at the front is torn. What's needed is a hefty dose of TLC so Mark and upholsterer Sonnaz Nooranvary put their heads together in collaboration.

Mark disassembles the radiogram to deal with the chassis and turntable first. He soon realizes that the chassis will need a lot of work. He removes a platter from another turntable, but there's an issue with the idler wheel he'll need to fix before repurposing the part. The motor needs monitoring with a voltmeter to see if can still rotate. The bearings are worn, he finds, and there is no grease around the mechanism. The good news is, however, that the motor is still functioning and can be made to work again. So, he focuses his attention on finding a solution for his idler wheel problem. He drops by to see horologist Steve Fletcher and asks him if he can adjust the idler wheel by making the inner circle a little larger.

Sonnaz has discovered that the crimson fabric the front of the radiogram is covered with wasn't manufactured from the mid-1960s onward. So she's found a new audio-friendly material as a replacement. A clever fix is required though; the new fabric is quite see-through, so she lines it with a red backing for authenticity and to complement the cream and gold of the cabinet.

Steve has done a fantastic job with the idler wheel. Mark devises a clever way to tackle the radio frequency finder and finishes the electronics. He's hoping to hear some sound; and he's pleased to hear noise emanating from both speakers.

Sisters Fredricka and Louisa return to the barn. They are so excited. They've brought with them a record of Mona Lisa by Nat King Cole. As the notes drift out across the barn, the sisters are so overcome, they can't help but cry happy tears.

Rachel's Go-Kart

A very special set of wheels

The bright yellow 'GeGe' electric go-kart brought into the barn by Emma Smith looks like a lot of fun, but it holds bittersweet memories of her childhood in Leeds and her older sister, Rachel, born with spina bifida. With spina bifida, a baby's spine and spinal cord do not develop properly. In Rachel's case, it was so severe that she couldn't walk or stand, spending a lot of time in a wheelchair, and she was always looking for ways to get about.

Her parents were dedicated to finding ways to improve her mobility and bought her the kart as a 10th birthday present. Dad Geoffrey had a custom-made metal panel installed to give Rachel's legs extra support, and replaced the footbrake with a handbrake.

Tragically, at Christmas in 1977, just a few days before her 10th birthday, Rachel became poorly. On New Year's Eve, her birthday, she died, before she even had a chance to sit in her go-kart. Heartbroken, her parents packed the kart away in the loft, along with all of Rachel's other unopened birthday presents – it was simply too painful to look at them.

Emma now has a family of her own; daughter Seren, aged ten, and son Milo, aged six, and her partner Dean. She has a vague memory of her dad eventually bringing the kart out of the loft many years later for Emma and her brother to play with. But, sadly, it didn't work.

In January 2021, the family suffered further heartbreak, when Emma's dad passed away. It was then decided to sell the family home and move her mum closer to her. When Emma was helping her mum, Bertie, pack, the final item that came down from the loft was the bright yellow kart. Emma's mum expressed how much she would like the kart to be restored so that it could be enjoyed by Seren and Milo, the niece and nephew Rachel never met. It would also create a special connection to Rachel and her dad for all the family. But the kart is now without a battery, and the specially adapted handbrake isn't connected.

Emma says she would like it to look just as it was, and wants to keep the number plate intact because it's a reminder of Rachel and the fact that it was bought for her. Such a job requires the expertise of electrical whizz Mark Stuckey – he's never worked on a go-kart before, he admits. He begins by stripping it down, to check the wiring and draws up a diagram to guide his repair. Although the kart has been stored in the loft for 40-odd years, the wiring is in very good condition. And the motor is still running, a great relief.

Next, he enlists the help of bicycle restorer Tim Gunn to help with switching back the specially-converted handbrake to a footbrake. There's quite a lot of superficial rust on the metal parts, so Mark uses wire wool to rub it down. The plastic cabin of the kart needs respraying

yellow, but the stickers have been there for so long, he fears if he tries to lift them they are likely to rip. So he protects them from the spray with tape. It would be so easy to go down the replacement line, he says, but the originality would be lost. Everything works and Mark is delighted as he puts everything back together.

Emma arrives back at the barn with Milo, Seren and her mum. Emma explains to Mark and Jay that it's really important to her for her mum to see it finished and fixed. She thinks that Mark has done an amazing job, and Bertie agrees. As Seren zooms off around the yard, her grandmother says she never expected to see this poignant Christmas present like this again. The last time she saw it was on the landing, the day her daughter Rachel died.

Emma has her wish come true, to have Rachel's go-kart looking as it did on the Christmas morning a few days before her sister died.

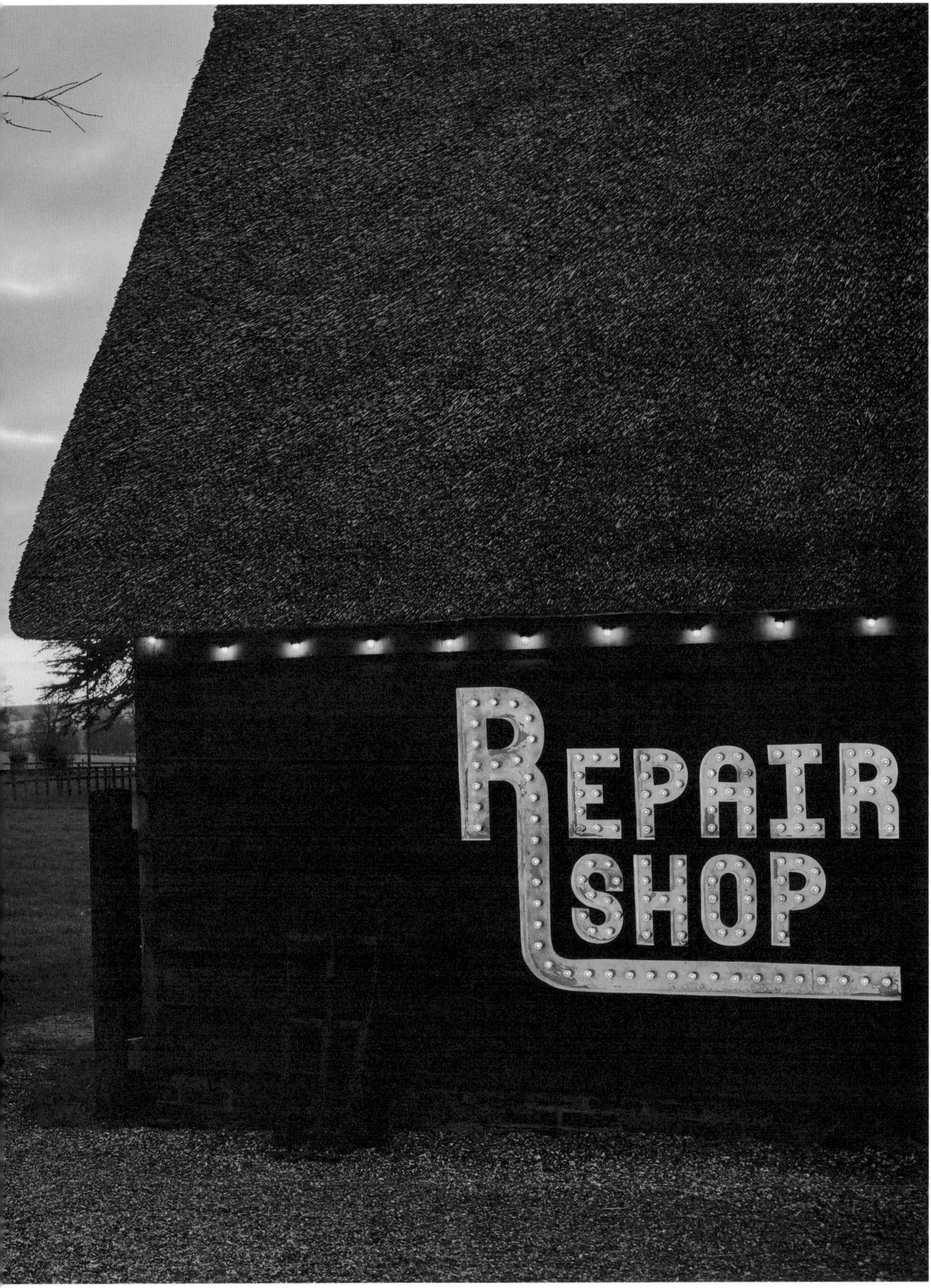
REPAIR
SHOP

Index

The Repair Shop

The chair that begins the story ...

This is Penny Young and this is her armchair. After Penny died, her daughter had the chair repaired out of respect and love for her Mum.

That daughter was Katy Thorogood, Ricochet's Creative Director and it's this chair that sparked the idea that became *The Repair Shop*.

Kyle Books would like to thank the amazing team at Ricochet for being so welcoming during our barn shoots. In particular:

Jo Ball, Emma Walsh, Alex Raw, Tanveer Bari,
Hannah Lamb, Martha Lefler, Sarah Wilson & Rich Merritt